PATHOLOGY AND LABORATORY MEDICINE

RATTUS NORVEGICUS

A REVIEW AND DIRECTIONS FOR RESEARCH

PATHOLOGY AND LABORATORY MEDICINE

Additional books and e-books in this series can be found on Nova's website under the Series tab.

PATHOLOGY AND LABORATORY MEDICINE

RATTUS NORVEGICUS

A REVIEW AND DIRECTIONS FOR RESEARCH

ANTONIETA MARIA ALVARADO MUÑOZ

AND

ANA ISABEL ROCHA FAUSTINO

EDITORS

nova
science publishers
New York

NOTICE TO THE READER

Library of Congress Cataloging-in-Publication Data

ISBN: 978-1-53614-685-1
LCCN: 2018963245

Published by Nova Science Publishers, Inc. † *New York*

We dedicate this book to our Families

CONTENTS

PREFACE

Since the human experimentation was abolished, the use of animal species similar to humans as much as possible has been imperative for a better understanding of human diseases, and for the development of novel prophylactic and therapeutic strategies to fight them.

Among all available species, *Rattus norvegicus* is considered an excellent species to study human and animal diseases. When choosing an animal for modeling a particular human disease, the researchers should take into account practical aspects, namely its facility of manipulation, number of animals needed for the experiment, its size and accommodation, its similarity with humans, among other factors specific for each experiment. When compared with other species, the rats meet the requisites of the researchers for the majority of the experiments.

In this way, countless works have been performed with rats along the last century, demonstrating the wonderful ability of this species to adapt to almost all circumstances that they are subjected to. Over years, the rats have been used to study distinct diseases, like cancer, diabetes, obesity cardiovascular and neurological diseases, among many others. Among all experiments using the rats as a model, two directions were followed in this book, putting in evidence the consolidated experience of three distinct research groups on the use of rats as model of neurological diseases and mammary carcinogenesis.

Neurological diseases in humans are considered as very important and relevant disorders by the World Health Organization, because of the impact that they may generate in the quality of life of human beings.

In the first part is presented the history and contemporary use of genetic-engineered rat models to study six important human neurological disorders: Autism, Parkinson's, Huntington's and Alzheimer's disease, and two neuromotor dysfunctions (Amyotrophic lateral sclerosis and Ataxia). This interesting review put in evidence the advantages of rat models when compared with other rodent models, like mouse models, for the study of these diseases, and present the model that best suits to the study of each disease.

The second part is focused on the importance of the rat as a model of oncological diseases, specifically on the study of mammary cancer. Mammary cancer remains as one of the most frequent cancers among women worldwide. Although several therapeutic approaches are available for the treatment of this disease, it is still responsible for the death of many women in the world every year. In this part, the authors presented the diversity of rat models available for the study of mammary cancer, with special emphasis on those chemically-induced, and showed the importance of the rat on the development of new prophylactic and therapeutic strategies to fight mammary cancer.

We are sure that the reading of this book will be very useful for those researchers aiming to use *Rattus norvegicus* as model of disease, providing them with valuable information for the selection and use of this species in their experiments.

In: *Rattus norvegicus*   ISBN: 978-1-53614-685-1
Editors: Antonieta Alvarado, et. al © 2019 Nova Science Publishers, Inc.

Chapter 1

WHY AREN'T THERE MORE GENETIC RAT MODELS OF NEUROLOGICAL DISEASES?

*Wesley M. Tierney, Aaron J. J. Lemus, Crystal T. Lao,
Bianca A. Ortega, Michelle Olmos, Jacqueline Saenz,
Toni L. Uhlendorf and Randy W. Cohen*
Department of Biology, California State University,
Northridge, CA US

ABSTRACT

Until recently, mice were the preferred mammal for the development of easy-to-produce, transgenic models of neurological diseases. Although rats are behaviorally, morphologically, and physiologically closer to humans, difficulties with pronuclei injection, efficiency in ova overproduction, and low implantation effectiveness restrict their application as models for neurological disorders. Recent advances in embryo-based targeting technology have generated several reliable, efficient and economical specific rat models of progressive human disorders. Hence, transgenic rat models of many neurological disorders should allow a more detailed evaluation of innovative clinical methodologies such as tissue engineering, designer drugs, stem cell implantation, and gene replacement therapy. This chapter will provide the basic history and the contemporary use of genetic rat models to study various human neurological diseases such as Autism, Parkinson's, Huntington's, Alzheimer's and two neuromotor dysfunctions (Amyotrophic Lateral Sclerosis and Ataxia).

Keywords: neurological disorders, rat, transgenic model

INTRODUCTION

Laboratory animal models have been established to mimic various types of human neurological diseases, and its related physiology and concomitant behavior. The development of both drug-induced and genetic models allows better understanding of disease progression that has permitted advances in surgical, material and pharmacological treatments. While certain drug or surgically induced models of neurologic disorders have their advantages, targeting specific genes allows a disease state that best mimics human pathophysiology and the accompanying symptoms. In this chapter, we present a history and relevance of genetically modified and mutated rat neuro-models, along with current information of their assets and limitations for the following disorders: Alzheimer's Disease, Amyotrophic Lateral Sclerosis (ALS), Ataxia, Autism, Huntington's Disease, and Parkinson's Disease. These disease models (specifically in rats) were characterized by all practices of genetic modification techniques, including point mutations, transgenic alterations, stem cell transplantations, knockout, and knockin models.

The advent of transgenic technology has allowed for the development of many animal models, including nematodes (like *Caenorhabditis elegans*), insects (*Drosophila melanogaster*), zebrafish, mice, and, of interest to this chapter, rats. While both mice and rats are genetically close to humans compared to the other frequently used invertebrate and fish models, mice have dominated the field of transgenic modeling since they were the first mammal to have well-developed techniques for gene manipulation (Sosa et al., 2012). Compared to mice, rats were more difficult to genetically modify twenty years ago. However, rat models have many advantages because they are biochemically, physiologically and morphologically closer to human brains than mice (Carmo and Cuello, 2013). The larger brain and neuron sizes in rats make it easier to perform neurosurgical and neuroimaging procedures. Furthermore, rats display more complex social behaviors than mice that can be attributed to the fact that, like humans, rats have significantly greater postnatal brain development. Given these advantages, competent genetic rat models have been a major goal for medical researchers, especially within the neurodegenerative spectrum. Yet, anticipation about creating and using genetic rat models for these diseases has yet to reach its zenith; a 2012 comprehensive review of all the genetic models of neurode-generative diseases from flies to rats listed only four rat models that had been created (Sosa et al., 2012). Certainly, much more effort should be done to correct the unavailability of the perfect model animal.

There are three technical difficulties that account for the scarcity of rat models. First, the creation of transgenic models requires many unfertilized ova which utilizes the practice of superovulation where the normal production of eggs in female rodents is manipulated to significantly increase the number of viable embryos (Popova et al., 2002). Generally, superovulation involves the overuse of various hormones such as Follicle Stimulating Hormone (FSH) or Pregnant Mare Serum Gonadotropin (PMSG). While neither of these hormones appears to affect survival rates of the ova, FSH alone seems to produce the highest amount of targeted eggs (Popova et al., 2002). However, more recent experiments showed that PMSG in coordination with another hormone, Human Chorionic Gonadotropin (HCG) was more successful at superovulation (Filipiak et al., 2006). However, even with the inclusion of multiple ovulation hormone treatments, the fact remains that rats still produce fewer viable ova compared to mice (Charreau et al., 1996; Tesson et al., 2005); this lack of embryo use has certainly stagnated further work and overall creation of transgenic rat models.

The second issue observed with rat transgenic neuro-models involves ineffective exogenous DNA microinjections and the subsequent decreased rate of survival of the zygotes and the ultimate creation of transgenic models. The process of microinjection involves injecting the foreign DNA (like human genes) into the rat fertilized embryo after superovulation, and then either culturing it first or immediately placing it back into the surrogate mother (Charrea et al., 1996). This extremely complicated procedure is made more difficult because of the more flexible and stickier cytoplasmic and pronuclear membranes of the rat embryos (Charreau et al., 1996; Popova et al., 2004; Tesson et al., 2005; Bugos et al., 2009). After superovulation, the survival rate of rat zygotes is 65% as compared to 80% in mice (Tesson et al., 2005), consequently after culture only about 35% - 61% of embryos are viable for microinjection (Tesson et al., 2005). Problematically, the success rate for rat transgenesis has been reported to be near 1% while mice have significantly better implantation rates between 10% - 20% (Popova et al., 2004; Tesson et al., 2005, Bugos et al., 2009).

Finally, the most recent yet elusive technique in creating transgenic rat models has been the use of embryonic stem cells (ES cells) to create novel knockout and knock-in disease models both *in vitro* and *in vivo*. This regenerative medicine procedure provides researchers with the full range of manipulating the embryos without the use of superovulation, microinjection and re-implantation and all the complications those entail. As far back as 1994, transgenic rat ES cells were isolated and genetically manipulated, but were reported to cause teratomas a few days after injection (Iannaccone et al., 1994). However, Iannaccone's cells did manage to differentiate into cells of all three germ layers. One reason for the lack of rat ES cells has been due to Oct-4 transcription factors, a key protein to induce pluripotency, which was discovered to be down regulated in rat blastocysts, limiting the critical creation of rat ES cells (Mullins et al., 2004). It wasn't until 2010 when competent rat ES cells were finally produced and used to make a p53 knockout rat (Tong et al., 2010). However, even with this worthwhile advancement in transgenic modeling, rat models are still not as numerous as mouse models and even less in neuro-disorders.

Given the difficulties that superovulation, microinjection and ES cell construction have in impeding the creation of rat neuro-models, mice have a near 30 year head start which has led to the plethora of transgenic mice models for neurodegenerative diseases. However, the advent of breakthrough of competent ES cells and recent molecular technologies like CRISPR/Cas9 should lead to the advent of many new rat models of neurodegenerative diseases. Additionally, where rat genes had been a challenge to manipulate, it has now become possible with advanced molecular techniques such as manipulating zinc finger nucleases to help insert exogenous DNA fragments (Gamber, 2016; Geurts et al., 2009) Ultimately, these new models should allow researchers to harness these advantages that rat models facilitate for testing treatments of various neurologic diseases.

HISTORY OF ANIMAL MODELS OF NEUROLOGICAL DISORDERS

Since the times of the ancient Greeks, animals have been utilized as comparative study subjects to model human physiology and anatomy (Franco, 2013). This practice allowed humans to conduct research while avoiding the social and ethical issues associated with the dissection of human cadavers (Staden, 1989); however, this did not come without dispute. Even centuries ago, vivisections raised alarms about whether animals were capable of feeling pain prior to death (Regan et al., 1989). Some philosophers deemed the dissection of animals to be immoral and unethical due to the unavailability of anesthesia (Franco, 2013). Other philosophers, like John Locke (1632-1704), argued against animal dissections for their potential in teaching children to be cruel to animals and humans alike (Locke and Quick, 1898). Nevertheless, the 18th century was accompanied by a new understanding about animals in research: animal suffering was permitted only if the research proved to be of greater significance for mankind (Boralevi et al., 1984). This more current philosophy shifted the main concern that overshadowed animal use in neurophysiological experiments from whether animals felt pain or not to whether the animal's suffering was justifiable in terms of the greater human good (Franco, 2013).

With the construction of new hospitals, training of doctors in medical schools, and the invention of improved surgical tools, the 19th century introduced greater change to the medical community (Brunton, 2004). In parallel with the medical revolution of the 19th century, animal research continued with the innovative end goal of developing therapies for pre-existing illnesses (Franco, 2013). In conjunction with a renewed purpose, anesthesia was introduced into animal model experimentation that allowed for animals to experience little or no pain during surgical procedures that was expected to result in postsurgical survival of the model (Cunningham et al., 1992). Still, some scientists rejected the use of anesthetics (like ether) entirely due to its inconsistent effects resulting from an incomplete understanding of dosage and concentration (Preece, 2011).

By the early 20th century, new vertebrate research models emerged that mitigated the public's concern over animal cruelty: the laboratory rat (Franco, 2013). In the 1800s, researchers began using an albino strain of the cosmopolitan brown rats (*Rattus norvegicus*) in many experiments. In an attempt to standardize rat experimentation, domestic rat strains were developed in the early 1900s with the Wistar Rat introduced as the first commercially available rat strain used in research in 1909 (Hedrich, 2000).

The 20th century was a problematic era in which two world wars were observed, followed by multiple economic recessions, and the growth of anti-vivisection activists, resulting in a lack of financial support for the usage of animals in laboratories around the world (Lederer, 1992; Liddick, 2006). However, rodents today are more accepted by the public for research due to being categorized as repugnant creatures, and additionally, they offer superior scientific benefits compared to other vertebrate preclinical models (Rudacille, 2001; Kaliste and Mering, 2007). The benefits of using rats as compared to mice are highlighted in current rat models and proposed drug, cell or surgical treatments of neurodegeneration (Thomas et al., 2003). Rats have significantly larger brains and therefore are more comparable to humans with regard to pharmacological testing, neurological sampling for biochemical analyses, and enhanced survival rates following microsurgeries for preclinical procedures such as stem cell implantations (Bugos et al., 2009). Rats are also valuable at various behavioral testing making them better subjects for various motor and learning disorders such as Autism Spectrum Disorder (Wohr and Scattoni, 2013).

Financially, compared to larger mammals and especially primates, rats are comparatively inexpensive to establish breeding colonies and are simple to manipulate both pharmacologically and surgically (Kaliste and Mering, 2007). Thus, with the availability of new laboratory rat strains, preclinical animal research has shifted to utilizing rodents in place of larger mammals such as dogs, pigs and horses (Baker et al., 1980; Kaliste and Mering, 2007). Consequently, today there are more than 200 strains of laboratory rats used as models of human diseases such as cancer, cardiovascular diseases, and of concern to the six neurologic disorders as described below, neurophysiological disorders. Despite all the complications limiting the number of transgenic models, there have been several rat neurodegenerative models created in the past 10 years.

With the construction of new hospitals, training of doctors in medical schools, and the invention of improved surgical tools, the 19th century introduced greater change to the medical community (Brunton, 2004). In parallel with the medical revolution of the 19th century, animal research continued with the innovative end goal of developing therapies for pre-existing illnesses (Franco, 2013). In conjunction with a renewed purpose, anesthesia was introduced into animal model experimentation that allowed for animals to experience little or no pain during surgical procedures that was expected to result in postsurgical survival of the model (Cunningham et al., 1992). Still, some scientists rejected the use of anesthetics (like ether) entirely due to its inconsistent effects resulting from an incomplete understanding of dosage and concentration (Preece, 2011).

By the early 20th century, new vertebrate research models emerged that mitigated the public's concern over animal cruelty: the laboratory rat (Franco, 2013). In the 1800s, researchers began using an albino strain of the cosmopolitan brown rats (*Rattus norvegicus*) in many experiments. In an attempt to standardize rat experimentation, domestic rat strains were developed in the early 1900s with the Wistar Rat introduced as the first commercially available rat strain used in research in 1909 (Hedrich, 2000).

The 20th century was a problematic era in which two world wars were observed, followed by multiple economic recessions, and the growth of anti-vivisection activists, resulting in a lack of financial support for the usage of animals in laboratories around the world (Lederer, 1992; Liddick, 2006). However, rodents today are more accepted by the public for research due to being categorized as repugnant creatures, and additionally, they offer scientific benefits compared to other vertebrate preclinical models (Rudacille, 2001; Kaliste and Mering, 2007). The benefits of using rats as compared to mice are highlighted in current rat models and proposed drug, cell or surgical treatments of neurodegen-eration (Thomas et al., 2003). Rats have significantly larger brains and therefore are more comparable to humans with regard to pharmacological testing, neurological sampling for biochemical analyses, and enhanced survival rates following microsurgeries for preclinical procedures such as stem cell implantations (Bugos et al., 2009). Rats are also valuable at various behavioral testing making them better subjects for various motor and learning disorders such as Autism Spectrum Disorder (Wohr and Scattoni, 2013).

Table 1. List of all cited rat disease models

Disease	Model	Use of Model	Mutation	Reference
Amyotrophic Lateral Sclerosis	G93A	Studies of age-dependent degeneration of motor neurons; characteristic of ALS	Overexpressing *SOD1*	Van Den Bosch, 2011
	H46R	Studies of age-dependent degeneration of motor neurons; characteristic of ALS	337	
	M337V	Studies of age-dependent degeneration of motor neurons; characteristic of ALS displaying degeneration of cerebellar, cortical, and hippocampal neurons	*TDP-43* expression	
Alzheimer's Disease	McGill-R-Thy1-APP	Increase expression of the single transgene in neurons to replicate AD pathology	*hAPP* cDNA with Swedish double mutation and Indiana mutation under the control of the murine thymocyte antigen promoter (*Thy1.2*)	Leon et al., 2010
	PSAPP or Tg478/Tg1116/Tg1587	PDGFβ promoter limits expression to neurons, especially in the hippocampus and cerebral cortex; Aβ plaques formed but genetic stress of triple transgene causes premature death	Expressed *hAPP* with Swedish and London (*K670N/M671L/V7*)	Flood et al., 2009
	Tg478/Tg1116	synapsin I promoter restricts expression to the brain and spinal cord; sufficient Aβ accumulation for plaque formation	Expressed *hAPP* with Swedish mutation	
	TgF344-AD	Neuroinflammation, gliosis, astrocytosis, with significant Aβ deposits	Express the *hAPP* gene with Swedish mutation and *PS1ΔE9* mutant gene	Cohen et al., 2013
	UKUR19	No Aβ accumulation	Human *PS1* gene with Finn mutation (*M671L*)	Kloskowska et al. 2010
	UKUR25	Intracellular Aβ accumulation in the hippocampus and cortex and signs of mild spatial memory impairment	Express human *PS1* gene with Finn mutation and *hAPP* gene with Swedish mutation	
	UKUR28	Intracellular Aβ accumulation in the hippocampus and cortex and signs of mild spatial memory impairment	Express *hAPP* with Swedish double mutation of *K670N, M671L*	

Disease	Model	Use of Model	Mutation	Reference
Ataxia	SCA3 (MJD)	Study of RNA interference-based therapeutics (SCA3)	Overexpression or silencing of wild-type *ataxin-3*	Paulson, 2011
	SCA17 (TBPQ64)	Study the assessment of different interventions on SCA17 phenotype	64 CAA/CAG repeats (TBP gene)	Kelp et al., 2013
	spastic Han-Wistar	Study of progressive cerebellar ataxia accompanied by loss of Purkinje cells; study of the use of stem cells in treating ataxia	Progressive death of Purkinje cells	Uhlendorf et al., 2011; Nuryyev et al., 2017
Autism Spectrum Disorder	FMR1 knockout	Memory deficits	Knockout of the *Fmr1* gene	Hamilton et al., 2014
	Methyl CpG binding protein 2	Behavioral impairments	Knockout of *MeCP2* gene	Nagarajan et al., 2006
	Neuroligin 3	Social impairments	*Neuroligin 3* knockout	Philibert et al., 2000
	NRXN1	Impairments in weight, sensory perception, learning, memory, attention, and social interactions	*Neurexin-1-alpha (NRX1a)* protein knockout	Esclassan et al., 2015
	SD FMR1	Displays elevated basal protein and increases in mGluR dependent long-term depression	Knockout of the *Fmr1* gene	Till et al., 2015
Huntington's Disease	BACHD	Study of memory deficits, movement, behavior, and food intake.	97 CAG repeats in *huntingtin* gene	Yu-Taeger et al., 2012
	tgHD	Study of progressive deterioration of memory, movement, and emotions; as well as decreases in body weight and dykinetic head movements	51 CAG repeats in *huntingtin* gene	von Hörsten et al., 2003
Parkinson's Disease	α-synuclein	Study normal function of the protein and aggregation in PD	*α-synuclein*; Autosomal-dominant	Lim and Ng, 2009; Blesa et al., 2012; Welchko et al., 2012
	DJ-1	Study role of protein in PD	*DJ-1*; Autosomal-recessive	Gamber, 2016
	LRRK2	Study role of protein in PD	*LRRK2*; Autosomal-dominant	Lim and Ng, 2009
	parkin	Study role of protein in PD	*parkin*; Autosomal-recessive	Dave et al., 2014
	PINK1	Study role of protein in PD	*PINK1*; Autosomal-recessive	Gamber, 2016

Financially, compared to larger mammals and especially primates, rats are comparatively inexpensive to establish breeding colonies and are simple to manipulate both pharmacologically and surgically (Kaliste and Mering, 2007). Thus, with the availability of new laboratory rat strains, preclinical animal research has shifted to utilizing rodents in place of larger mammals such as dogs, pigs and horses (Baker et al., 1980; Kaliste and Mering, 2007). Consequently, today there are more than 200 strains of laboratory rats used as models of human diseases such as cancer, cardiovascular diseases, and of concern to the six neurologic disorders as described below, neurophysiological disorders. Despite all the complications limiting the number of transgenic models, there have been several rat neurodegenerative models created in the past 10 years (Table 1).

AMYOTROPHIC LATERAL SCLEROSIS (ALS)

Amyotrophic lateral sclerosis (ALS; commonly referred to as Lou Gehrig's disease) is a devastating neurological disease characterized by selective degeneration of motor neurons found within the motor cortex and alpha-motor neurons within the brain and spinal cord, respectively (Lee et al., 2013; Thonhoff et al., 2007). This selective degeneration of motor neurons causes skeletal muscle dysfunction among individuals with ALS, eventually leading to paralysis and ultimately death (within 2-6 years after diagnosis) usually due to failure of respiratory muscles (Lee et al., 2013; Lepore et al., 2011; Picher-Martel et al., 2016). The causes of ALS can be divided into two categories. First, familial ALS (fALS) is caused by a genetic origin (like *superoxide dismutase*) in roughly 20% of cases (Andjus et al., 2009). Alternatively, most cases of ALS are of the sporadic type (sALS) where the cause is still unknown (Andjus et al., 2009).

According to Picher-Martel et al. (2016), rats are of particular interest for ALS preclinical studies because of their larger brain and spinal cord size. In addition, their outsized neuron morphology (compared to mice) allows researchers to perform complex pharmacological procedures, such as intrathecal or intracerebroventricular injections for preclinical drug testing (Picher-Martel et al., 2016). As a result of pathological and clinical likenesses to human fALS, rat models have been established using hemizygous transgenic rats that express the specific mutated *human superoxide dismutase 1* (*Cu/Zn-SOD1*) gene (Thonhoff et al., 2007). These transgenic rats represent the preeminent model of preclinical ALS studies compared to ALS mice (Damme et al., 2017). Additionally, these rat models provided significant insights into mammalian motor neuron degeneration pathways found in ALS (Damme et al., 2017). In the Van Den Bosch (2011) review of genetic rodent models of ALS, he explained how transgenic rats overexpressing mutant *SOD1* (*H46R* and *G93A* knock-ins) were produced using genetic modifications, and that these rats developed an age-dependent degeneration of motor neurons, causing progressive paralysis and death. Additionally, a benefit of this ALS model allowed biochemical analyses of the spinal cord and spinal fluid, using implantation of intrathecal catheters for progressive therapeutic studies and clinical testing of cerebrospinal fluid. Also discussed was an additional ALS rat model that was created with mutant *TDP-43* (M337V), a mutated protein expressed only in ALS-affected neurons.

This model was produced by expressing the mutated gene under the mechanism of a promoter that can be easily silenced by administering the transgenic rats with doxycycline. The *M337V* rat model presented ALS-like neuronal degeneration of spinal motor, cerebellar, cortical and hippocampal neurons. Experiments with this rat ALS model indicated that it replicated some but not all the pathogenic characteristics observed in ALS patients. In summary, the mutant *SOD1* rat models as described above remain the best comparative models currently available to study the pathogenesis of ALS, allowing the examination of new treatments and basic neurophysiology. Nevertheless, the issue remains how representative these models are for the more common ALS (Van Den

Bosch, 2011). Obviously, due to this limitation, there is a great need for improved genetic rat ALS models.

ALZHEIMER'S DISEASE (AD)

Alzheimer's disease (AD) is an incurable neurological disorder that affects about 15% of people over 65 years of age and is the most common cause of dementia in the United States (Hebert et al., 2013). The characteristic symptoms of AD are progressive neuronal degenerations linked to an irreversible decline of cognitive function, including memory retention and spatial orientation. The AD brain shows atrophy directly related to the loss of neurons concentrated in the cerebral cortex and hippocampus. One of the most studied aspects of AD is the formation of extracellular amyloid-β (Aβ) plaques. The amyloid precursor protein (APP) is cleaved by β- and γ-secretases, resulting in altered Aβ peptides that clump to produce the anatomical hallmark of AD, the plaques. Another distinct feature of AD is the abnormal accumulation of neurofibrillary tangles (NFTs). Abnormal phosphorylation or hyperphosphorylation of the Tau protein creates paired helical filaments that disrupt the structural integrity of axonal microtubules, leading to the formation of NFTs (Joo et al., 2017). The combined neurodegenerative abilities of Aβ plaques and NFTs is accompanied by neuroinflammation from abnormal activation of astrocytes and microglial cells, assisting ultimately to observed neurodegeneration (Crews et al., 2010; Reisberg et al., 1987).

The earliest transgenic rat models for AD were first established by Echeverria et al. (2004). Three transgenic rat lines were developed from the Wistar strain with the intention of inducing extracellular Aβ accumulation in cortical and hippocampal pyramidal neurons. These models (coded as *UKUR28*) had the transgene for human APP (*hAPP*) containing both the Swedish double mutations (*K670N, M671L*) and the Indiana mutation (*V717F*) (Echeverria et al., 2004). However, the

UKUR28 rat models did not produce extracellular amyloid plaques, they did show intracellular Aβ accumulation in cortical and hippocampal neurons and displayed some signs of spatial memory impairment (Kloskowska et al., 2010). Based on previous experiments with transgenic rodents, Echeverria et al. (2004) suggested that the moderate levels of APP transgene expression were not sufficient to result in Aβ plaque formation in these transgenic rats, concluding that APP overexpression would be required for a better AD rat model.

The first transgenic rat models to successfully develop amyloid plaques were generated in 2009, but NFTs were absent. Both double (*Tg478/Tg1116*) and triple transgenic rat models (*PSAPP* or *Tg478/Tg1116/Tg11587*) were generated by crossing lines of Sprague-Dawley rats. The *Tg478* line expressed hAPP with the Swedish mutation under the control of a rat synapsin I promoter that restricted expression to the brain and spinal cord. The *Tg1116* line expressed hAPP containing the Swedish and London (*K670N/M671L/V717I*) mutations under the control of the *PDGFβ* promoter, which limits expression to neurons in the hippocampus and cerebral cortex. The *Tg11587* line expressed hPS1 with the Swedish/Finnish (*M146V*) mutation under the control of a rat synapsin I promoter (Flood et al., 2009). The double homozygous AD rats exhibited sufficient Aβ production for extracellular deposition. Although this transgenic rat model exhibited progressive accumulation of Aβ plaques with some activation of astrocytes and microglia, concomitant behavior deficits, however, neuronal degeneration was absent. Furthermore, this model was prone to premature death due to hypertension, immunosuppression, and kidney disease caused by genetic disturbance as a result of the triple transgenes (Carmo and Cuello, 2013; Klowskowska et al., 2010).

The *McGill-R-Thy1-APP* transgenic rat line was developed in 2010 from Wistar rats (Leon et al., 2010). The DNA construct used for the pronuclear infection carried hAPP complementary DNA (cDNA) with the Swedish double mutation and the Indiana mutation under the control of the murine thymocyte antigen promoter (*Thy1.2*). The researchers sought to increase expression of the single transgene in neurons to replicate AD pathology (Leon et al., 2010). Intracellular and extracellular Aβ plaques were observed in the cortex and hippocampus with recruitment of activated microglia, inducing the associated neuroinflammation surrounding the plaques along with cognitive impairment. The *McGill-R-Thy1-APP* rat line has been used as a model for many studies involving AD. For example, metabolite levels in the cortex and hippocampus were studied using magnetic resonance spectroscopy, and these findings demonstrated complex progression of AD metabolites in *McGill-R-Thy1-APP* rats not observed in aging control rats (Nilsen et al., 2012), correlating to the human condition.

The first transgenic rat model to encompass all the anatomical and behavioral hallmarks of AD was the *TgF344-AD* line. Fischer-344 rats were manipulated to express the *APP* gene with the Swedish mutation and the Δ exon 9 mutant *PS1 gene* (*PS1ΔE9*) under the control of a mouse prion promoter (Cohen et al., 2013). *TgF344-AD* rats demonstrated progressive accumulation of intraneuronal Aβ, soluble and insoluble Aβ40 and Aβ42 peptides, and amyloid plaques. The transgenic rats also showed neuroinflammation, gliosis, astrocytosis, and significant Aβ deposits that were observed throughout the brain. The *TgF344-AD* rats also developed NFT pathology independent of any human tau mutations, relying solely on the endogenous rat tau protein (Cohen et al., 2013). Amyloidosis and tauopathy were accompanied by concomitant pyramidal neuron death in the cortex and hippocampus. Cognitive performance was observed in neurological reflex tests, open field test, object recognition tests, and the Barnes maze assay. *TgF344-AD* rats exhibited significantly impaired learning and memory, declining with age. The *TgF344-AD* model is capable of producing the full spectrum of AD pathologies and correlated behavioral deficits, generating preclinical experiments that should elucidate the cause and effect mechanisms of AD leading to innovative safe and viable treatments.

ATAXIA

Ataxia is a neurological disease that is defined by the lack of coordination of muscle movement, behavioral abnormalities, and progressive mental impairment (Urbach et al., 2010). Ataxias come in many forms affecting a variety of cerebellar and spinal cord neurons associated with movement. According to Urbach et al., (2010), there are several probable reasons for the many forms of ataxia. Patients experiencing hereditary ataxic condition are challenged with progressive neurodegeneration of the cerebellum, particularly histopathologic variations from the norm and significant loss of Purkinje cells (Uhlendorf et al., 2011). Analyses of human and rat studies have shown critical biochemical changes in the neurons associated with ataxia, leading to cerebellar and spinal cord deterioration (Uhlendorf et al., 2011).

The spastic Han-Wistar (sHW) serves as an exceptional rat model to explore treatments that reduce ataxic mobility. This autosomal recessive rat model experiences progressive cerebellar ataxia complete with a steady loss of Purkinje cells, beginning at 30 days of age. In our laboratory, this rat model has been used to analyze drugs and certain physical therapies such as the positive impact of chronic exercise on slowing Purkinje cell neurodegeneration (Uhlendorf et al., 2011). Recently, Nuryyev et al. (2017) showed that human neural progenitor cells (hNPCs) elicited signs of repair and restoration of motor function in this ataxic rat. Results of this study proved that stem cells have a future clinical use correcting ataxia and related neuro-motor disorders (Nuryyev et al., 2017).

Other hereditary ataxias are autosomal, dominantly-inherited neurodegenerative dysfunctions developed from CAG/polyglutamine repeat sequences (Urbach et al., 2010). Researchers have utilized two of the following rat models for genetic ataxias: a transgenic rat model of spinocerebellar ataxia type 17 (*SCA17*) and a virus-based rat model of spinocerebellar ataxia type 3 (*SCA3*) (Urbach et al., 2010). The rat model of *SCA17* confers a complete length of human cDNA construct that encodes for the mutant TATA-box binding proteins (Urbach et al., 2010). Moreover, the rat model of SCA3 has been used in studies for Machado-Joseph disease (MJD) (Urbach et al., 2010). Specifically, in a study done by Paulson (2011), this rat model of MJD endured either overexpressing or silencing of wild-type *ataxin-3*. Results from this study showed that a therapeutic approach involving non-allele-specific silencing to treat MJD patients was both successful and safe (Paulson, 2011).

Kelp et al. (2013) studied the *SCA17* rat model that conveys a full human cDNA fragment of the TBP gene with 64 CAA/CAG repeats (*TBPQ64*). Through their study of the *TBPQ64* rats, Kelp et al. (2013) confirmed that diffusion tensor imaging (DTI) can be utilized to explore the cerebellum for neuropathological modifications, searching for possible diagnostic biologic markers for *SCA17* patients (Kelp et al., 2013). These studies have demonstrated that transgenic rat models resemble the human state of ataxia, hence making them the preferred model to use in research studies to discover improved treatments for patients.

AUTISM SPECTRUM DISORDER (ASD)

Autism Spectrum Disorder (ASD), according to the Center for Disease Control and Prevention (CDC) survey of parents, is prevalent in 1 in 45 children in the USA, approximately 2% (CDC, 2015); currently, an estimated 1.5 million adults are living with autism (CDC, 2007). ASD is a disorder encompassing a broad spectrum with varying divergent symptoms. The core symptoms include social deficits, language impairments, and intellectual disabilities (Seltzer et al., 2003). Other neurological ASD symptoms can include sleep problems, mood changes, anxiety, and hyperactivity. A related disorder and possibly an onset version of ASD is Fragile X Syndrome (FXS) with symptoms mirroring that of ASD such as intellectual disability, problems with social interaction, and delayed speech (Crawford et al., 2001). The rat ASD model offers an excellent experimental choice because of the stronger social and complex behavioral aspects like social play (Siviy and Panksepp, 2011). However, due to the complications of genetically manipulating rats, very few ASD and FXS rat models have been established so far. As better and more efficient transgenesis methods for rats have been discovered, more models of these disorders will be created for use, mirroring the human symptoms of ASD and FXS.

Due likely to the lack of genetic understanding of ASD in humans or because of a better understanding of a direct genetic cause of FXS, more FXS rat models have been created. FXS is thought to be caused by a triple CGG repeat in the *fragile X mental retardation 1* (*FMR1*) gene which in turn leads to a decreased expression of the *fragile X mental retardation protein* (*FMRP*) (Crawford et al., 2001). *In situ, FMRP* regulates synapses between neurons in humans. While this is the common patho-physiology of FXS, this poor connectivity among neurons may correlate to the overconnectivity hypothesis about ASD, thus connecting the two disorders symptomatically and physiologically (Courchesne et al., 2007).

Intelligence and memory are both affected in FXS and ASD humans. FXS rat models have similar dysfunctions in memory deficits stemming from abnormal protein expression and the resultant morphological changes within the brain. The FXS rat model (*SD FMR1*) was developed from a knockout transgene. The *SD FMR1* model displayed higher synthesis of the mutated protein leading to greater amounts of metabo-tropic glutamate receptor (mGluR) linked synapses in CA1 pyramidal neurons in hippocampus (Till et al., 2015). Associated memory, the main function of the hippocampus, was found to be significantly reduced within these mutants while spatial learning was determined to be unaffected (Till et al., 2015). Physiologically, these rats were found to have larger spine head diameters in hippocampal mGluR1-associated dendritic spines in early postnatal ages. When FMR1 knockouts were tested for memory deficits, researchers revealed significant deficits as compared to normal rats (Hamilton et al., 2014).

Behavior impairments are inherent with autistic humans manifested in social deficiencies and repetitive behavior (Seltzer et al., 2003). This form of behavior is also observed in many models of ASD and FXS. *Neuroligin 3* (*NLGN3*) is a gene associated with ASD and is down-regulated in humans with ASD, leading to impairments in social inter-actions (Philibert et al., 2000). Both *FMR1* and *NLGN3* knockouts were tested for various social behaviors, and were both found to have insufficiencies in social interactions compared to normal rats (Hamilton et al., 2014). Interestingly, some social behaviors, including physical contact and sniffing, were found to be higher when compared with normal rats (Hamilton et al., 2014). Finally, *methyl CpG binding protein 2* (*MeCP2*) is a gene that encodes the *MeCP2* protein (found in epigenetic targeting of neurons) and was found in lower levels in humans with ASD as well as in *NLGN3* knockout rats (Nagarajan et al., 2006), suggesting an epigenetic link.

Lastly, *NRXN1* is a gene that encodes the *Neurexin-1-alpha* (*NRX1a*) protein that has been shown to aid in presynapse establishment (Li et al., 2006). Related to this subchapter, the *NRX1a* protein was found to be altered in humans with ASD and other neurocognitive disorders (Südhof, 2008). A Sprague-Dawley rat knockout of this gene (referred to as *NRXH1a*) was created and tested for various ASD-related behaviors (Esclassan et al., 2015). The *NRXH1a* rats were found to have impair-ments in weight, sensory perception, learning, memory, attention, and social interactions. Interestingly, while male and female *NRXH1a* rats had equal hyperactivity, the male knockout rats were more likely to be impaired compared to the female knockout rats, corroborating the observation that human males are four times as likely to have *NRX1a* impairments than females (Esclassan et al., 2015). Finally, *NRXH1a* rats exhibited supersensitivity to sound similar to humans with ASD. The researchers concluded that the *NRX1a* knockout was a good candidate for a rat model of ASD (Esclassan et al., 2015).

HUNTINGTON'S DISEASE (HD)

Huntington's disease (HD) is a fatal, autosomal dominant neurode-generative disease that leads to deterioration of healthy nerve cells primarily in the neostriatum and cerebral cortex. The common symptoms of HD are the slow decline of cognitive, behavioral and motor functions. The disease often presents in people starting at 45-55 years of age, but early onset as young as 2 years old and as late as 80 years has been observed. Eventually, the HD motor-degenerating symptoms lead to death approximately 15-20 years after the first symptoms are diagnosed (Roos et al., 1993). HD roughly affects 10 to 15 people per 100,000 in Europe (Praveen and Roger, 2015); HD has the same inheritance rate in men and women (Evans et al., 2013).

The cause of Huntington's disease is a mutation in *huntingtin* gene (*HTT*) that codes for the *huntingtin* protein which performs such tasks as chemical signaling, protein binding, and apoptosis protection in the neostriatum. In HD, *HTT* mutations cause expanded CAG trinucleotide repeats on the DNA, leading to elongated segments of *huntingtin* protein. The unusually long *huntingtin* protein is cut into smaller detrimental fragments that bind and alter the function of neostriatal neurons, leading to cell death correlated with the onset of HD symptoms (Imarisio et al., 2008).

The first transgenic rat model of HD (tgHD) was completed by von Hörsten et al. (2003). A rat model was preferred because the mouse models have multiple restrictions of a smaller neuron size and less efficient learning and memory system compared to rats. This HD model has 51 CAG repeats in their mutated *huntingtin* gene that led to slow progressively disease state. The model experienced HD symptoms such as a decrease in body weight and increased dyskinetic movements in the head. The tgHD model also slowly developed emotional, cognitive, and motor deteriorations during their life spans analogous to HD patients. While the researchers have been able to produce models with similar HD motor and cognitive symptoms, they were not able to mimic the HD progressive effects on the rat over time (von Hörsten et al., 2003). Upon closer inspection, the tgHD rat lacked the full-length mutant *HTT* protein found in humans.

Another rat model with symptoms of HD is the *bacterial artificial chromosome HD* (BACHD) transgenic rat model expressing the full-length mutant *HTT* protein (97 CAG repeats). This model shows advantages over the *tgHD* rat with faster progressive deterioration in motor impairments and anxiety (Yu-Taeger et al., 2012). Additionally, this model was tested for its memory, operational motor behavior and mood. Similar to human symptoms that varied depending on the number of repeated sequences, the BACHD rat model verified that the model illustrated HD symptoms, including neostriatal neuropathology and concomitant cognitive impairments (Clemensson et al., 2017). BACHD rats also displayed changes in metabolism (another symptom in HD patients) with less motivation to eat compared to normal rats. Behavioral tests were also accomplished and correlated with the progressive decline in phenotypic symptoms (Jansson et al., 2014). Other researchers have confirmed that the BACHD model displays evidence of progressive cognitive and motor deficits that are similar to HD patients (Fielding et al., 2012). The low number of affected HD patients could be a financial reason why the paucity of transgenic rat HD models (Vlamings et al., 2012).

PARKINSON'S DISEASE (PD)

Demographic trends in the aging population around the world have shown that Parkinson's disease (PD) is the second most common neurodegenerative disease of the elderly (Blesa et al., 2012). PD is characterized by the degeneration of dopaminergic (DA) neurons in the substantia nigra and the formation of protein aggregates known as Lewy bodies (Tieu, 2011). The progressive degeneration of DA neurons leads to a gradual deterioration of motor function. At the beginning, this deterioration in movement can be characterized by tremor, limb rigidity, slow movement, gait and balance problems. While the phenotypic cause of PD is due to lack of dopamine regulation in basal ganglia, the exact reason why PD causes its numerous clinical effects remains muddled. To date, the knowledge obtained about dopaminergic function in motor control has helped in analyzing anatomical connections as well as the neuropathological and pharmacological changes related to PD (Van Kampen et al., 2017). Current treatments such as physical therapy, pacemakers, laser therapy and drugs such as catechol-o-methyl-transfer-ase inhibitors, dopamine agonists, and levodopa only treat the symptoms and have no real effect on reducing neuronal degeneration in PD patients (Jankovic and Aguilar, 2008; Gamber, 2016). Due to its prevalence in the growing population, it is imperative to seek more understanding on the cause and effects of PD with the use of non-drug induced rat models (Lim and Ng, 2009).

As discussed in the Introduction, rats are the preferred animal model for neurodegenerative diseases, and transgenic models have a significantly higher validity than the neurotoxin rodent models (i.e., MPTP and 6-OHDA). Induced cell death in the substantia nigra by neurotoxins does not yield reliable models because these rats do not offer the ability to study PD progression and the resultant neuropathology that occurs. Over a decade ago, the rat genome was sequenced giving a greater opportunity to understand and eventually model human disease, allowing the development for novel drug treatments that show a more translational approach to study PD. Genetic mutations in PD only account for 10% of cases (Blesa, 2012). The genes that have been linked to familial PD include α-*synuclein* (α-*syn*), *leucine-rich repeat serine /threonine kinase 2* (*LRRK2*), *PTEN-induced putative kinase 1* (*PINK1*), *parkin* and *DJ-1* (Lim and Ng, 2009) and have been used in many mice models. These genetic defects have also been instructive in building rat models with insights into the mechanisms of PD expression as well as the accompanying pathogenesis.

The first gene α-*synuclein (α-syn)* to be associated with Parkinson's Disease also helped produce the most thoroughly studied genetic rat model of PD (Welchko et al., 2012). From its initial link to PD, the α-*syn* model paved the way to identify several other candidate genes allowing for PD research advancement. Mutations in the α-*syn* gene are characterized by the formation of Lewy bodies, an anatomical hallmark of α-*syn* misfolding. SAGE Labs developed a rat model using the CRISPR/Cas9 genome targeting strategy with the research application of dopaminergic cell toxicity. Previous research indicated that α-*syn* knockout had no effect on the DA neurons in mice, counteracting those results in the *Drosophila* that displayed DA cell loss (Blesa et al., 2012). Like α-*syn*, studies on *LRRK2* also showed no signs of DA cell loss in mice models. The knockout rat model developed by SAGE Labs indicated a complete loss of the target protein leading to neuronal apoptosis of DA neurons in the substantia nigra. This model has made it ideal to study PD (Lim and Ng, 2009).

The other three genes *PINK1*, *Parkin*, and *DJ-1* all appear to play a role in the phenotypic effects observed familial parkinsonism (similar to PD). It has been noted that the mutations in these genes do not actually cause PD, but instead leads to the related parkinsonism syndrome (Ahlskog, 2009). The *Parkin* knockout model has been used to study the early onset of PD by assessing dopaminergic neurotoxicity (Dave et al., 2014). One problem with this PD model is normal motor activity on the rotarod. Loss of motor function is a common characteristic in PD, but this lack of motor performance loss is yet again a limiting factor for this particular PD model. The *PINK1* model used the knockout of this gene to result in complete loss of the protein and was characterized by the hind limb dragging, fatigue and foot slips (Gamber, 2016). In addition, preliminary results indicated approximately 50% DA neuron reduction in the substantia nigra. Finally, another genetic knockout rat model *DJ-1* exhibited complete loss of the protein and was characterized by the hind limb dragging, gait impairments, and an observed 50% decrease of DA neurons in the substantia nigra (Gamber, 2016). *PINK1* and *DJ-1* rats were the first genetic models to express progressive loss of DA neurons and corresponding moderate to severe motor deficits (Gamber, 2016). The symptom spectrum that these rat models present is encouraging for the biomedical field of PD research.

To this end, many models have been developed in the attempt to find better therapies for PD patients. As far as biomedical research, no perfect rat model has been able to successfully express the full spectrum of progressive characteristics of PD as yet. However, studying the many rat models should allow for a more translational approach displaying a larger spectrum of PD phenotypes compared to the mouse, allowing researchers to study the progression and pathology of Parkinson's disease to develop better therapies.

CONCLUSION

While mice have been favored for genetic model creation due to ease and genetic efficiency, genetic rat models are superior for many neurological diseases (especially the ones listed in this chapter) and are on the rise. Motor, behavioral, and memory dysfunctions are significantly better models with rats compared to mice, highlighting the greater need for more rat models of the diseases. Fortunately, due to advances in genetic modeling, researchers have been able to create some rat models of major neurodegenerative diseases. These new and future models take advantage of the superior symptom modeling of the rats and the capacity to examine more treatments *in vivo*.

REFERENCES

Ahlskog, E. J. (2009). *Parkin* and *PINK1* parkinsonism may represent nigral mitochondrial cytopathies distinct from Lewy body

Parkinson's disease. *Parkinsonism and Related Disorders*, *15*, 721-27.

Andjus, P. R., Bataveljić, D., Vanhoutte, G., Mitrecic, D., Pizzolante, F., Nevena, D., Nicaise, N., Gankam, F., Gangitano, C., Michetti, F., van der Linden, A., Pochet, R. & Bacić, G. (2009). *In vivo* morphological changes in animal models of Amyotrophic Lateral Sclerosis and Alzheimer's-like disease: MRI approach. *The Anatomical Record*, *292*, 1882-92.

Baker, H. J. J., Lindsey, R. & Weisbroth, S. H. (2013). *The laboratory rat: biology and diseases.*, Vol. 1. Elsevier.

Blesa, J., Phani, S., Jackson-Lewis, V. & Przedborski, S. (2012). Classic and new animal models of Parkinson's disease. *BioMed Research International*, *12*, 845618.

Boralevi, L. C. (1984). *Bentham and the oppressed*, Vol. 1. Walter de Gruyter.

Brunton, D. (2004). *Medicine transformed: health, disease and society in Europe 1800-1930.* Manchester University Press.

Bugos, O., Bhide, M. & Zilka, N. (2009). Beyond the rat models of human neurodegenerative disorders. *Cellular and Molecular Neurobiology*, *29*, 859-69.

Center for Disease Control (CDC). (2007). *CDC releases new data on autism spectrum disorders (ASDs) from multiple communities in the United States*, Centers for Disease Control and prevention release, http://www.cdc.gov/od/oc/media/pressrel/2007/r070208. htm.

Center for Disease Control. (2015). *National health interview survey*. https://www.cdc.gov/nchs/data/nhsr/nhsr087.pdf.

Charreau, B., Tesson, L., Soulillou, J. P., Pourcel, C. & Anegon, I. (1996). Transgenesis in rats: technical aspects and models. *Transgenic Research*, *5*, 223-34.

Clemensson, E., Håkan, K., Clemensson, L. E., Riess, O. & Nguyen, H. P. (2017). The *BACHD* rat model of Huntington disease shows signs of fronto-striatal dysfunction in two operant conditioning tests of short-term memory. *PLoS One*, https://doi.org/10.1371/journal.pone.0169051.

Cohen, R. M., Rezai-Zadeh, K., Weitz, T. M., Rentsendorj, A., Gate, D., Spivak, I., Bholat, Y., Vasilevko, V., Glabe, C. G., Breunig, J. J., Rakic, P., Davtyan, H., Agadjanyan, M. G., Kepe, V., Barrio, J. R., Bannykh, S., Szekely, C. A., Pechnick, R. N. & Town, T. (2013). A transgenic Alzheimer rat with plaques, tau pathology, behavioral impairment, oligomeric aβ, and frank neuronal loss. *Journal of Neuroscience, 33*, 6245-56.

Courchesne, E., Pierce, K., Schumann, C. M., Redcay, E., Buckwalter, J. A., Kennedy, D. P. & Morgan, J. (2007). Mapping early brain development in Autism. *Neuron, 56*, 399-413.

Crawford, D. C., Acuña, J. M. & Sherman, S. L. (2001). *FMR1* and the *Fragile X* syndrome: human genome epidemiology review. *Genetics in Medicine, 3*, 35-71.

Crews, L., Rockenstein, E. & Masliah, E. (2010). APP Transgenic modeling of Alzheimer's disease: mechanisms of neurodegeneration and aberrant neurogenesis. *Brain Structure and Function, 214*, 111-26.

Cunningham, A. & Williams, P. (1992). *Anaesthetics, ethics and aesthetics: vivisection in the late nineteenth-century British laboratory. The laboratory revolution in medicine.* New York: Cambridge University Press.

Dave, K. D., Silva, S., Sheth, N. P., Ramboz, S., Beck, M. J., Quang, C., Switzer, R. C., Ahmad, S. O., Sunkin, S. M., Walker, D., Cui, X., Fisher, D. A., McCoy, A. M., Gamber, K., Ding, X., Goldberg, M. S., Benkovic, S. A., Haupt, M., Baptista, M. A., Fiske, B. K., Sherer, T. B. & Frasier, M. A. (2014). Phenotypic characterization of recessive gene knockout rat models of Parkinson's disease. *Neurobiology of Disease, 70*, 190-203.

Dayalu, P. & Albin, R. L. (2015). Huntington disease: pathogenesis and treatment. *Neurologic Clinics, 33*, 101-14.

Do Carmo, S. & Cuello, A. C. (2013). Modeling Alzheimer's disease in transgenic rats. *Molecular Neurodegeneration*, 8, 37.

Echeverria, V., Ducatenzeiler, A., Alhonen, L., Janne, J., Grant, S. M., Wandosell, F., Muro, A., Baralle, F., Li, H., Duff, K., Szyf, M. & Cuello, A. C. (2004). Rat transgenic models with a phenotype of intracellular Aβ accumulation in hippocampus and cortex. *Journal of Alzheimer's Disease*, 6, 209-19.

Esclassan, F., Francois, J., Phillips, K. G., Loomis, S. & Gilmour, G. (2015). Phenotypic characterization of nonsocial behavioral impairment in *Neurexin 1α* knockout rats. *Behavioral Neuroscience*, 129, 74-85.

Evans, S. J. W., Douglas, I., Rawlins, M. D., Wexler, N. S., Tabrizi, S. J. & Smeeth, L. (2013). Prevalence of adult Huntington's disease in the UK based on diagnoses recorded in general practice records. *Journal Neurology, Neurosurgery, and Psychiatry*, 84, 1156-60.

Fielding, S. A., Brooks, S. P., Klein, A., Bayram-Weston, Z., Jones, L. & Dunnett, S. B. (2012). Profiles of motor and cognitive impairment in the transgenic rat model of Huntington's disease. *Brain Research Bulletin*, 88, 223-36.

Filipiak, W. E. & Saunders, T. L. (2006). Advances in transgenic rat production. *Transgenic Research*, 15, 673-686.

Flood, D. G., Lin, Y. G., Lang, D. M., Trusko, S. P., Hirsch, J. D., Savage, M. J., Scott, R. W. & Howland, D. S. (2009). A transgenic rat model of Alzheimer's disease with extracellular Aβ deposition. *Neurobiology of Aging*, 30, 1078-90.

Franco, N. H. (2013). Animal experiments in biomedical research: a historical perspective. *Animals*, 3, 238-73.

Gamber, K. M. (2016). Animal models of Parkinson's disease: new models provide greater translational and predictive value. *BioTechniques*, 61, 210-11.

Geurts, A. M., Cost, G., Freyvert, Y., Zeitler, B., Miller, J. & Choi, V. (2009). Knockout rats produced using designed zinc finger nucleases. *Science*, 325(5939), 433.

Hamilton, S. M., Green, J. R., Veeraragavan, S., Yuva, L., McCoy, A., Wu, Y., Warren, J., Little, L., Ji, D., Cui, X., Weinstein, E. & Paylor, R. (2014). *Fmr1* and *Nlgn3* knockout rats: novel tools for investigating autism spectrum disorders. *Behavioral Neuroscience, 128*, 103-9.

Hebert, L. E., Weuve, J; Scherr, P. A. & Evans, D. A. (2013). Alzheimer disease in the United States (2010–2050) estimated using the 2010 census. *Neurology, 80*, 1778-83.

Hedrich, H. J. (2000). *The history and development of the rat as a laboratory animal model. The laboratory rat.* Stein: Switzerland.

Iannaccone, P. M., Taborn, G. U., Garton, R. L., Caplice, M. D. & Brenin, D. R. (1994). Pluripotent embryonic stem cells from the rat are capable of producing chimeras. *Developmental Biology, 163*, 288-92.

Imarisio, S., Carmichael, J., Korolchuk, V., Chen, C. W., Saiki, S., Rose, C., Krishna, G., Davies, J. E., Ttofi, E., Underwood, B. R. & Rubinsztein, D. C. (2008). Huntington's disease: from pathology and genetics to potential therapies. *Biochemical Journal, 412*, 191-209.

Jankovic, J. & Aguilar, L. G. (2008). Current approaches to the treatment of Parkinson's disease. *Neuropsychiatric Disease and Treatment, 4*, 743-57.

Jansson, E., Clemens, L. E., Riess, O. & Nguyen, H. P. (2014). Reduced motivation in the BACHD rat model of Huntington disease is dependent on the choice of food deprivation strategy. *PLoS One*, doi:10.1371/journal.pone.0105662.

Joo, I. L., Lai, A. Y., Bazzigaluppi, P., Koletar, M. M., Dorr, A., Brown, M. E., Thomason, L. A. M., Sled, J. G., McLaurin, J. & Stefanovic, B. (2017). Early neurovascular dysfunction in a transgenic rat model of Alzheimer's disease. *Scientific Reports* doi: 10.1038/srep46427.

Kaliste, E. (2004). *The welfare of laboratory animals*, Vol. *2*. Springer Science & Business Media.

Kelp, A., Koeppen, A. H., Petrasch-Parwez, E., Calaminus, C., Bauer, C., Portal, E., Yu-Taeger, L., Pichler, B., Bauer, P., Riess, O. &

Nguyen, H. P. (2013). A novel transgenic rat model for spinocerebellar ataxia type 17 recapitulates neuropathological changes and supplies *in vivo* imaging biomarkers. *Journal of Neuroscience, 33,* 9068-81.

Kloskowska, E., Pham, T. M., Nilsson, T., Zhu, S., Öberg, J., Codita, A. & Pedersen, L. A. (2010). Cognitive impairment in the *Tg6590* transgenic rat model of Alzheimer's disease. *Journal of Cellular and Molecular Medicine, 14,* 1816-23.

Lederer, S. E. (1992). Political animals: the shaping of biomedical research literature in twentieth-century America. *Isis, 83,* 61-79.

Lee, J. D., Kamaruzaman, N. A., Fung, J. N. T., Taylor, S. M., Turner, B. J., Atkin, J. D., Woodruff, T. M. & Noakes, P. G. (2013). Dysregulation of the complement cascade in the *hSOD1 G93A* transgenic mouse model of Amyotrophic Lateral Sclerosis. *Journal of Neuroinflammation, 26*(10), 119.

Leon, W. C., Canneva, F., Partridge, V., Allard, S., Ferretti, M. T., DeWilde, A., Vercauteren, F., Atifeh, R., Ducatenzeiler, A., Klein, W., Szyf, M., Alhonen, L. & Cuello, A. C. (2010). A novel transgenic rat model with a full Alzheimer's-Like Amyloid pathology displays pre-plaque intracellular amyloid-β associated cognitive impairment. *Journal of Alzheimer's Disease, 20,* 113-26.

Lepore, A. C., O'Donnell, J., Kim, A. S., Williams, T., Tuteja, A., Rao, M. S., Kelley, L. L., Campanelli, J. T. & Maragakis, N. J. (2011). Human glial-restricted progenitor transplantation into cervical spinal cord of the *SOD1G93A* mouse model of ALS. *PloS One,* https://doi.org/10.1371/journal.pone.0025968.

Li, X., Zhang, J., Cao, Z., Wu, J. & Shi, Y. (2006). Solution structure of *GOPC PDZ* domain and its interaction with the c-terminal motif of neuroligin. *Protein Science, 15,* 2149-58.

Lim, K. L. & Ng, C. H. (2009). Genetic models of Parkinson disease. *Biochimica Et Biophysica Acta (BBA)-Molecular Basis of Disease, 1792,* 604-15.

Locke, J. (1712). *Some thoughts concerning education. Modern history sourcebook: John Locke (1632-1704)* A. & J. Churchill.

Mullins, L. J., Wilmut, I. & Mullins, J. J. (2004). Nuclear transfer in rodents. *The Journal of Physiology, 554*, 4-12.

Nagarajan, R., Hogart, A., Gwye, Y., Martin, M. R. & LaSalle, J. M. (2006). Reduced *MeCP2* expression is frequent in autism frontal cortex and correlates with aberrant *MeCP2* promoter methylation. *Epigenetics, 1*, 172-82.

Nilsen, L. H., Melø, T. M., Sæther, O., Witter, M. P. & Sonnewald, U. (2012). Altered neurochemical profile in the *McGill-R-Thy1-APP* rat model of Alzheimer's disease: a longitudinal *in vivo* 1H MRS study. *Journal of Neurochemistry, 123*, 532-41.

Nuryyev, R. L., Uhlendorf, T. L., Tierney, W., Zatikyan, S., Kopyov, O., Kopyov, A., Ochoa, J., Van Trigt, W., Malone, C. S. & Cohen, R. W. (2017). Transplantation of human neural progenitor cells reveals structural and functional improvements in the spastic Han-Wistar rat model of ataxia. *Cell Transplantation, 26*, 1811-21.

Paulson, H. (2012). Machado-Joseph Disease/spinocerebellar staxia sype 3. *Handbook of Clinical Neurology, 103*, 437-49.

Philibert, R. A., Winfield, S. L., Sandhu, H. K., Martin, B. M. & Ginns, E. I. (2000). The structure and expression of the human *neuroligin-3* gene. *Gene, 246*, 303-10.

Picher-Martel, V., Valdmanis, P. N., Gould, P. V., Julien, J. P. & Dupré, N. (2016). From animal models to human disease: a genetic approach for personalized medicine in ALS. *Acta Neuropathologica Communications, 4*(1), 70.

Popova, E., Krivokharchenko, A., Ganten, D. & Bader, M. (2002). Comparison between PMSG- and FSH-induced superovulation for the generation of transgenic rats. *Molecular Reproduction and Development: Incorporating Gamete Research, 63*, 177-82.

Popova, E., Krivokharchenko, A., Ganten, D. & Bader, M. (2004). Efficiency of transgenic rat production is independent of transgene-construct and overnight embryo culture. *Theriogenology*, *61*, 1441-53.

Preece, R. (2011). *The history of animal ethics in western culture. The psychology of the human-animal bond.* Springer, New York, New York.

Regan, T. (1980). Utilitarianism, vegetarianism, and animal rights. *Philosophy & Public Affairs.* Princeton University Press.

Reisberg, B., Borenstein, J., Salob, S. P. & Ferris, S. H. (1987). Behavioral symptoms in Alzheimer's disease: phenomenology and treatment. *The Journal of Clinical Psychiatry*, *48*, 9-15.

Roos, R. A., Hermans, J., Der Vlis, M., Van Ommen, G. J. & Bruyn, G. W. (1993). Duration of illness in Huntington's disease is not related to age at onset. *Journal of Neurology, Neurosurgery & Psychiatry*, *56*, 98-100.

Rudacille, D. (2000). *The scalpel and the butterfly: the war between animal research and animal protection.* Macmillan.

Seltzer, M. M., Krauss, M. W., Shattuck, P. T., Orsmond, G., Swe, A. & Lord, C. (2003). The symptoms of Autism Spectrum disorders in adolescence and adulthood. *Journal of Autism and Developmental Disorders*, *33*, 565-81.

Siviy, S. M. & Panksepp, J. (2011). In search of the neurobiological substrates for social playfulness in mammalian brains. *Neuroscience & Biobehavioral Reviews*, *35*, 1821-30.

Sosa, M. A. G., De Gasperi, R. & Elder, G. A. (2012). Modeling human neurodegenerative diseases in transgenic systems. *Human Genetics*, *131*, 535-63.

Südhof, T. C. (2008). Neuroligins and neurexins link synaptic function to cognitive disease. *Nature*, *455*, 903-11.

Tesson, L., Cozzi, J., Menoret, S., Remy, S., Usal, C., Fraichard, A. & Anegon, I. (2005). Transgenic modifications of the rat genome. *Transgenic Research*, *14*, 531-46.

Thomas, M. A., Chen, C. H., Jensen-Seaman, M. I., Tonellato, P. J. & Twigger, S. N. (2003). Phylogenetics of rat inbred strains. *Mammalian Genome, 14*, 61-64.

Thonhoff, J. R., Jordan, P. M., Karam, J. R., Bassett, B. L. & Wu, P. (2007). Identification of early disease progression in an ALS rat model. *Neuroscience Letters, 415*, 264-68.

Tieu, K. (2011). A guide to neurotoxic animal models of Parkinson's disease. *Cold Spring Harbor Perspectives in Medicine, 1*(1), a009316.

Till, S. M., Asiminas, A., Jackson, A. D., Katsanevaki, D., Barnes, S. A., Osterweil, E. K., Bear, M. F., Chattarji, S., Wood, E. R., Wyllie, D. J. & Kind, P. C. (2015). Conserved hippocampal cellular pathophysiology but distinct behavioural deficits in a new rat model of FXS. *Human Molecular Genetics, 24*, 5977-84.

Tong, C., Li, P., Wu, N. L., Yan, Y. & Ying, Q. L. (2010). Production of *p53* gene knockout rats by homologous recombination in embryonic stem cells. *Nature, 467*, 211-13.

Uhlendorf, T. L., Van Kummer, B. H., Yaspelkis, B. B. & Cohen, R. W. (2011). Neuroprotective effects of moderate aerobic exercise on the spastic Han–Wistar Rat, a model of Ataxia. *Brain Research, 1369*, 216-22.

Urbach, Y. K., Bode, F. J., Nguyen, H. P., Riess, O. & Hörsten, S. (2010). Neurobehavioral Tests in rat models of degenerative brain diseases. *Rat Genomics, 597*, 333-56.

Van Damme, P., Robberecht, W. & Van Den Bosch, L. (2017). Modelling Amyotrophic Lateral Sclerosis: progress and possibilities. *Disease Models and Mechanisms, 10*, 537-49.

Van Den Bosch, L. (2011). Genetic rodent models of amyotrophic lateral sclerosis. *BioMed Biotechnology Research International*, 2011, 348765.

Van Kampen, J. M. & Robertson, H. A. (2017). The BSSG rat model of Parkinson's disease: progressing towards a valid, predictive model of disease. *EPMA Journal, 8*, 261-71.

Vlamings, R., Zeef, D. H., Janssen, M. L. F., Oosterloo, M., Schaper, F., Jahanshahi, A. & Temel, Y. (2012). Lessons learned from the transgenic Huntington's disease rats. *Neural Plasticity*, 2012, 682712.

von Hörsten, S., Schmitt, I., Nguyen, H. P., Holzmann, C., Schmidt, T., Walther, T. & Bader, M. (2003). Transgenic rat model of Huntington's disease. *Human Molecular Genetics*, *12*, 617-24.

Von Staden, H. & Chalcedonius, H. (1989). *Herophilus: the art of medicine in early Alexandria: edition, translation and essays.* Cambridge University Press.

Welchko, R. M., Lévêque, X. T. & Dunbar, G. L. (2012). Genetic rat models of Parkinson's disease. *Parkinson's Disease*, *33*, 717-29.

Wöhr, M. & Scattoni, M. L. (2013). Behavioural methods used in rodent models of autism spectrum disorders: current standards and new developments. *Behavioural Brain Research*, *251*, 5-17.

Yu-Taeger, L., Petrasch-Parwez, E., Osmand, A. P., Redensek, A., Metzger, S., Clemens, L. E., Park, L., Howland, D., Calaminus, C., Gu, X., Pichler, B., Yang, X. W., Riess, O. & Nguyen, H. P. (2012). A novel *BACHD* transgenic rat exhibits characteristic neuropathological features of Huntington disease. *Journal of Neuroscience*, *32*, 15426-38.

In: *Rattus norvegicus*　　　　　　ISBN: 978-1-53614-685-1
Editors: Antonieta Alvarado, et al.　© 2019 Nova Science Publishers, Inc.

Chapter 2

THE RELEVANCE OF *RATTUS NORVEGICUS* IN MAMMARY CANCER RESEARCH

Antonieta Alvarado[1,2,*], Ana I. Faustino-Rocha[1,2,*],
Bruno Colaço[1,3] and Paula A. Oliveira[1,4]
[1]Center for the Research and Technology of Agro-Environmental
and Biological Sciences (CITAB), UTAD, Vila Real, Portugal
[2]Faculty of Veterinary Medicine, Lusophone University of
Humanities and Technologies (ULHT), Lisbon, Portugal
[3]Department of Zootechnics, University of Trás-os-Montes
and Alto Douro (UTAD), Vila Real, Portugal
[4]Department of Veterinary Sciences, UTAD, Vila Real, Portugal

ABSTRACT

Mammary cancer in rats resembles that of women in several features, namely in its hormone responsiveness, histology, biochemical properties, molecular and genetic characteristics. Different strains of Rattus norvegicus have been used in the last years as an important tool for understanding many aspects of mammary cancer, and for the development

---

* These authors contributed equally to this work.

and evaluation of new preventive and therapeutic strategies for this disease. As mammary cancer is one of the most common cancers, victimizing more than half a million of women worldwide every year, the importance of the rat model of mammary cancer is increasing over years. Despite several studies in this field, the current therapeutic approaches, like surgery, radiotherapy, chemotherapy and immunotherapy are not effective, and some of them have devastating effects for patients. With this chapter, the authors intended to provide the readers with an overview of the rat models used to study mammary cancer, with a special emphasis on chemically-induced models.

Keywords: chemical carcinogenesis, mammary cancer, rodent model

INTRODUCTION

Cancer is one of the most important public health problems world-wide. Despite all the advances in the early diagnosis and treatment of this disease, it was considered the second leading cause of death globally by the World Health Organization (WHO) (Herbst et al., 2006; World Health Organization (WHO), 2018). According to the WHO, cancer was responsible for 8.8 million deaths in 2015, and globally, nearly 1 in 6 deaths were due to the cancer. Consequently, the economic impact of cancer is significant and is increasing over years. In 2010, the cancer costs approximately $ 1.16 trillion (World Health Organization (WHO), 2018). These data reinforces the idea that much remains to be done in cancer research.

Breast cancer remains as one of the most frequent cancers among women, affecting approximately one out of ten women worldwide (Liska et al., 2000). According to the World Health Organization, in the year 2015, breast cancer was responsible for the death of approximately 571 000 women around the world (World Health Organization (WHO), 2018). In this way, a better knowledge about the cancer biopathology is essential to the development of new preventive and therapeutic strategies that may improve the quality of life and lifespan of oncologic patients (Faustino-Rocha, 2017).

Since early times, animals have been used as models by researchers in order to disclose the anatomy and physiology of the human body. Aristotle (384-322 B.C.), who is considered one of the most important thinkers, used animals to study intern differences among species. His studies were well documented and spread to other countries, contributing to the use of animal models as a research tool in several European and Arabian countries (Ericsson, Crim, and Franklin, 2013). After this, animals have been frequently used and they have had a great impact in biomedical research. Proofing this is the fact that over the last century, all Nobel prizes in the field of physiology and medicine used animals as models of diseases (Badyal and Desai, 2014). Currently, animal models still represent an essential tool to study several diseases, including cancer. They allow the researchers not only to study the etiology, the mechanisms and the progression of this disease, but also to search for new therapeutic strategies that may improve the quality of life and lifespan of oncologic patients (Iannaccone and Jacob, 2009a).

ANIMAL MODELS: HOW TO SELECT THE CORRECT SPECIES?

Although several animal species, like fishes, rabbits, rats, mice, dogs, non-human primates and large animals are available as models, the researchers should be able to choose the most adequate species to their studies (Conn, 2013). An ideal animal model of human diseases should be simple, not expensive and similar to Human as much as possible (Fagundes and Taha, 2004). In order to choose the models that best fits to their studies, the researchers should consider the following aspects: the aim of the study, available species, advantages and disadvantages of each species, accommodation expenses, manipulation, required equipment and ethical considerations (Fagundes and Taha, 2004; Van der Gulden, Beynen, and Hau, 1999). Although several species are available, mice (*Mus musculus*) and rats (*Rattus norvegicus*) remains as the species more frequently used in research protocols performed in the European Union (Comission, 2013). When compared with other species, the mice and rats have some advantages, such as their use is easily approved by legislation on the protection of animals used for scientific purposes, their physiology and genetic are well known, they are relatively cheap, they are small animals, easy to accommodate and manipulate, and the most important one, they are mammals and have many similarities with humans, like anatomy, physiology, genetic and biochemistry (Committee for the Update of the Guide for the Care Animals, 2011; Kararli, 1995).

WHY USE *RATTUS NORVEGICUS* AS A MAMMARY CANCER MODEL?

Rat (*Rattus norvegicus*) is frequently used as model to the study of mammary cancer (Cardiff, 2007). The development of mammary cancer in rats was firstly reported by Mceuen in 1938, after the daily vaginal application of a solution of estrone in corn oil for two years and an half (Mceuen, 1938). The mammary cancer in rats resembles that of women in its hormone responsiveness, histology, biochemical properties, molecular and genetic characteristics. The use of this model has allowed a better understanding of many aspects of mammary carcinogenesis, like its genetic and molecular basis, pathogenesis, and the development and evaluation of several therapeutic approaches (Clarke, 1996; Faustino-Rocha et al., 2015b; Iannaccone and Jacob, 2009b; Liska et al., 2000). Moreover, the rats provide a higher quantity of blood and tissue samples for posterior studies, when compared with mice (Hoenerhoff et al., 2011; Mullins, Brooker, and Mullins, 2002).

ANATOMICAL AND HISTOLOGICAL FEATURES OF RAT MAMMARY GLAND

The female rat has two mammary chains (right and left) with six mammary glands with a nipple each one: three pairs in thoracic region (extended to the cervical region) and three pairs in abdominal-inguinal region (Maeda, Ohkura, and Tsukamura, 2000). The mammary glands of each mammary chain are usually numbered by the nipple from one to six in the cranio-caudal direction and the gland tissue of the thoracic region is smaller when compared with that of the abdominal-inguinal region (Hvid et al., 2011). Contrarily to that happens in women, the rat mammary glands are poorly developed and they may be identified externally only by the presence of the nipple. Rat mammary glands are greatly vascularized by the branches of several arteries, likey thoracic internal and external, superficial cervical, external pudendal and axillary (Maeda et al., 2000).

The rat mammary gland has a tubuloalveolar conformation, being composed by a group of branched tubular ducts and alveolar buds (AB) (Lucas et al., 2007). Histologically, it is mainly composed of two tissues: parenchymal and stromal tissue (connective, adipose and vascular network tissues) (Nandi, Guzman, and Yang, 1995). The rat mammary gland tissues have a higher stromal and parenchymal component, when compared with mice mammary gland (Hoenerhoff et al., 2011; Mullins et al., 2002). Similar to women, the rat mammary gland tissue is hormone dependent and grows during estrous cycle and pregnancy (Hvid et al., 2012).

The rats' mammary gland develops through different phases. An extensive development occurs during puberty by the day 21, being characterized by the differentiation of the epithelium into terminal end bud units that correspond to the bulbous end of the branch lactiferous duct (Colditz and Frazier, 1995; Lucas et al., 2007). The duct is composed by a layer of luminal epithelial cells (ductal epithelial cells) fated to form the walls of the ductal lumen with outer of myoepithelial cells and the basement membrane. Terminal end bud is composed by multi-layered of preluminal epithelial cells (also called body cells) surrounded by a layer of pluripotent stem cells (also called cap cells) that are progenitors of mioephitelial cells. Both body cells and cap cells are very proliferative (Cardiff, 2007; Hinck and Silberstein, 2005; Manivannan and Nelson, 2012). Terminal end bud proliferates dichotomously into alveolar buds and terminal ductules with continuous branching ducts and ductules that drain into the duct of the nipple (Hvid et al., 2012). Each alveolar buds and duct have one layer of simple epithelial cells surrounded by a layer of myoepithelial cells and the basement membrane, supported by the stroma. Near to 50-55 days of age, the alveolar lobules are formed from the alveolar buds in both non-pregnant and pregnant rats (Hvid et al., 2012; Lucas et al., 2007).

As the rat mammary gland tissue is hormone dependent, the secretory structure is influenced by the sexual maturity (reached at about 6 weeks of age), estrous cycle and pregnancy (Sengupta, 2013). Estrogen and growth hormone are the hormones mainly responsible for the ductal elongation. Despite these, other hormones like progesterone and thyroid hormones may influence the proliferation, branching and differentiation of mammary gland. In pregnancy, the progesterone is the main responsible for the exponential growth of the mammary gland, and the prolactin is responsible for the mammary gland alveologenesis and the development of specialized epithelium for milk production (Radisky, Hirai, and Bissell, 2003).

DEVELOPMENT OF MAMMARY CARCINOGENESIS IN RATS

In a general point of view, carcinogenesis is a multistep process progressing over many years. It consists of distinct but related phases in which different molecular and cellular alterations occur, this phases can be describe as follow:

- *Initiation* is characterized by the spontaneous or induced irreversible DNA-damage, leading to the conversion of a normal cell into an initiated one.

- *Promotion* is considered a relatively lengthy and reversible process during which the initiated cells grow and divide in an uncontrolled way as a result of accumulated abnormalities, originating a population of preneoplastic cells. The promotion may be changed by the administration of chemopreventive agents that can affect the tumor growth rates.

- *Progression* is characterized by a fast increase in tumor size. During this phase the preneoplastic cells may convert into neoplastic ones as a consequence of additional genetic alterations, and the tumor may become malignant and possibly metastatic.

- *Metastization* is a complex process during which the cancer cells spread from the primary tumor to discontiguous organs, through blood or lymphatic system. Although the ability to metastasize is exclusive of the malignant neoplasms, not all malignant neoplasms metastasize (Eickmeyer et al., 2012; Oliveira, 2016; Oliveira et al., 2006a, 2007; Siddiqui et al., 2015).

The faster expansion of the mammary gland epithelium that occurs during the rat puberty (45-55 days of age) is considered the key-point for the initiation of carcinogenesis (Sternlicht, 2006). The rat mammary tumors may emerge from epithelial cells located in duct, terminal end bud and alveolar buds. However, due to the high mitotic/proliferation index of pluripotent cells from terminal end bud (cell cycle takes approximately 13 hours) that conduct to the ductal elongation, ramification and cell differentiation, the terminal end bud is the most common site of mammary carcinogenesis initiation in female rats (Sternlicht, 2006). As it is commonly known, the pluripotent cells also called stem cells are very susceptible to DNA changes resulting in the easy formation of preneoplastic and neoplastic cells, which give rise to the primary tumor and potential generation of metastatic cells (Eickmeyer et al., 2012). As the cell cycle in terminal end bud is shorter when compared with alveolar buds, the time to repair the DNA damages is shorter, contributing to the higher susceptibility of terminal end bud for carcinogenesis initiation (Colditz and Frazier, 1995).

RAT STRAINS IN MAMMARY CARCINOGENESIS

Although several rat strains are available for the study of mammary carcinogenesis, they have different susceptibility for mammary tumors' development due to their different genetic background (Luzhna, Kutanzi, and Kovalchuk, 2015). The strains of the Norway rats (*Rattus norvegicus*) are the most frequently used in experimental assays (Maeda et al., 2000). The outbred strains of Sprague-Dawley and Wistar female rats are more sensitive to carcinogenic agent when compared with other strains, namely inbred August and Marshall (Boyland and Sydnor, 1962). Recent studies on chemically-induced mammary carcinogenesis verified that carcinogen-exposed female Wistar rats developed lower number of mammary tumors when compared with female Sprague-Dawley rats, suggesting that the last strain is more sensitive to carcinogen exposition (Gal et al., 2011). Although inbred female Fisher 334 rats have been widely used as a model of mammary carcinogenesis, the Sprague-Dawley and Wistar rats are the most frequently used strains due to the higher susceptibility of mammary tissue to the initiation of carcino-genesis after carcinogen exposition (Rudmann et al., 2012; Russo and Russo, 1996). Moreover, immunocompromised strains, like nude rat strain (rnu/rnu rat) may be used in the study of mammary carcinogenesis. This nude rat that has an autosomal recessive mutation known as *rnu* was backcrossed with different strains and consequently a high number of congenic strains characterized by hairlessness and congenital absence of thymus were generated. Due to the absence of cell-mediated immunity (immunocompromised animals), these strains are usually used as xenograft models in which human breast cancer cells are transplanted into rats (Marchesi, 2013; Rudmann et al., 2012; Schuurman, Hougen, and Van Loveren, 1992). The ACI strain that is the result of a cross between August and Copenhagen Irish rats was identified several years ago as a sensitive strain for mammary tumors development after a combined estrogen administration and radiation exposition (Holtzman, Stone, and Shellabarger, 1982; Shellabarger, Stone, and Holtzman, 1983). Because of the estrogens susceptibility and morphological similarities with human mammary tumors, the ACI strain is considered an ideal model to study estrogen-induced human breast carcinomas (Ravoori et al., 2007).

RAT MODELS OF MAMMARY CARCINOGENESIS

The extensive experimental research in the field of mammary carcinogenesis over years has conducted to the development of several rat models. Distinct approaches involving rats for modelling mammary cancer such as the induction of mammary tumors development by the administration of carcinogenic substances, irradiation or hormone administration, implantation of cancer cells, or use of genetically engineered animals have been established (Bartstra et al., 1998; Russo and Russo, 1996; Shull et al., 1997; Smits et al., 2007).

SPONTANEOUS TUMORS

The use of rat strains that spontaneously develop mammary tumors have an important role in the design of experimental protocols. Although the rat mammary gland is the second organ more frequently affected by the development of spontaneous neoplasms after the pituitary gland, it is age-dependent occurring mainly after the first year of age (Ikezaki, Takagi, and Tamura, 2011; Mcmartin et al., 1992; Oishi et al., 1995; Son et al., 2010). As the experimental protocols usually occurs during relatively short periods of time (before the first year of age of animals), the development of spontaneous mammary tumors will not interfere with the interpretation of the effects of a carcinogen agent. The mammary fibroadenoma (frequency from 18.9 to 57.0%), followed by adenocarcinoma (frequency from 8.8 to 16.8%) and adenoma (3.5 to 7.0%) are the spontaneous mammary neoplasms more frequently identified in female Sprague-Dawley rats (Chandra, Riley, and Johnson, 1992; Kaspareit and Rittinghausen, 1999; Mcmartin et al., 1992). Our research team has performed some studies in the field mammary carcinogenesis using the model of mammary cancer *N*-methyl-*N*-nitrosourea (MNU)-induced in female Sprague-Dawley rats. Unexpectedly, in the last experiment, a high-grade undifferentiated mammary carcinoma was identified in a seven-week-old female rat belonging to the control group of the experiment (the animal did not receive any drug) (Faustino-Rocha et al., 2016c, 2016a). To our knowledge, no previous reports had described a spontaneous mammary tumor in such a young rat.

INDUCED TUMORS

Although the rat spontaneous mammary tumors are uncommon, their development may be easily induced in immunocompetent rats through the administration of chemical carcinogens, hormone administration or exposure to physical agents (Russo and Russo, 1996).

CHEMICAL INDUCTION

A carcinogenic is any compound able to induce cancer development in living tissues (Oliveira et al., 2007). The first studies on chemically-induced mammary tumors in rats were described by Howell in 1963 (Howell, 1963). In experimental assays using animals for the study of chemical carcinogenesis, a simple, relatively fast and safe method for the administration of the carcinogenic agent is required (Arcos, 1995; Oliveira et al., 2006b).

The induction of rat mammary tumors by carcinogenic agents may be hormone dependent or independent. In the case of the hormone dependent mammary tumors, the first step of the carcinogenesis (initiation) greatly depends on the age of animals at the time of the chemical carcinogen administration (Nandi et al., 1995; Russo and Russo, 1996; Thordarson et al., 2001). The chemical carcinogenesis is maximal when the carcinogen agent is administered between 45 and 60 days of the age (animals in sexual maturity), which coincides with the time of active differentiation of terminal end bud to alveolar buds (Russo and Russo, 1996). It was also previously described that the ovariectomy after the chemical carcinogen administration may reduce the incidence of mammary tumors in 96%. However, although hormone dependent mammary tumors depend on estrogen hormones in an initial phase of the carcinogenesis, later they may progress to a more aggressive phenotype without estrogen dependency and expression of the respective receptors (Thordarson et al., 2001).

An adequate rat model of mammary carcinogenesis should exhibit histopathological features and genetic alterations similar to those described in women mammary tumors, the initial or intermediate lesions (preneoplastic lesions) should simulate the different steps of the carcino-genesis, the tumors should originate specifically from the mammary gland tissue, a higher incidence (higher than 60%) should be obtained in a relatively short period of time (latency period lower than six months), and the assay should be reproducible (Barros et al., 2004; Noble and Cutts, 1959; Russo and Russo, 1996; Steele and Lubet, 2010; Wagner, 2004). Until now, only two chemical carcinogenic agents for mammary carcinogenesis with these characteristics are well established: 12-dimethylbenz(a)-anthracene (DMBA) and MNU (Medina, 2007).

The DMBA is a classical polycyclic aromatic hydrocarbon commonly used to the induction of mammary tumors in rodents (Al-Dhaheri et al., 2008; Cortés-García et al., 2009; Currier et al., 2005; Russo and Russo, 2000). When is orally administered at 50-56 days of age, in a single dose ranging from 10 to 100 mg per kg of body weight, it induces the development of a high number of mammary tumors (Al-Dhaheri et al., 2008). To exert its carcinogenic effects, the DMBA must be previously bioactivated by the cytochrome P-450/P1-450 monooxygenase enzyme systems by hepatic way. The metabolites are mono- and dihydroxymethyl and they can also be metabolized to their corresponding dihydrodiols, phenols and other oxidation products (Russo, Tay, and Russo, 1982). The generated epoxides interact with the DNA generating the transversions A:T for T:A and G:C for T:A, through two mechanisms: the formation of adducts DMBA-adenine and DMBA-guanine, and the loss of purines by spontaneous lysis of the complex between the DMBA epoxide and DNA purines target (Cortés-García et al., 2009).

Inversely to the DMBA, the MNU is a direct alkylating agent that does not require the metabolic activation in order to induce irreversible changes in DNA (Doctores et al., 1974; Murray et al., 2009). It acts as carcinogen by methylation of the guanine nucleosides, promoting the GGA to GAA transitional mutation in H-ras codon 12 encoding glutamic acid in place of glycine (Lu et al., 1998; Murray et al., 2009; Sukumar et al., 1983). A single intraperitoneal (i.p.) injection of this carcinogen agent at 50 days of age, at a standard dose of 50 mg per kg of body weight induces mammary tumors in all susceptible animals (100% of incidence) (Alvarado et al., 2016; Faustino-Rocha et al., 2016b, 2015c; Murray et al., 2009). The MNU may also be administered by other ways, such as subcutaneous (s.c.), intravenous (i.v.) and oral (p.o.). However, the induction rate of mammary tumors by MNU depends on administration route and dose, being higher when MNU is intraperitoneally injected (Lu et al., 1998).

HORMONE INDUCTION

The development and maintenance of rat mammary tissue is hormone-dependent (Sengupta, 2013). Estrogens, which are involved in the normal development of mammary gland, have the ability to induce a tumorigenic response when administered in combination with chemical carcinogens or exposure to physical carcinogenic agents like irradiation (Russo and Russo, 1998).

The epithelial mammary tissue from ACI rats is particularly susceptible to estrogen administration, a proliferation and transformation of the epithelial cells from mammary gland is observed after the exposition to this hormone (Ravoori et al., 2007). Shull et al., performed an experimental assay with this strain in which 17β-estradiol pellets were implanted into ovary-intact ACI rats inducing mammary tumors development (mainly comedo carcinomas) in all exposed animals with a mean latency of 145 days (Shull et al., 1997). The susceptibility of mammary gland to other steroid hormones, such as 2-hydroxyestradiol, 4-hydroxyestradiol, 16-hydroxyestradiol, or 4-hydroxyestrone was different, with no induction of mammary tumors development when these steroid hormones were administered at a similar dose of 17β-estradiol (Turan et al., 2004).

PHYSICAL INDUCTION

The first studies describing the susceptibility of the Human mammary gland to ionizing radiation (gamma rays) of Hiroshima and Nagasaki atomic bombs clearly demonstrated the carcinogenic effects of this kind of radiation. Moreover, these studies also demonstrated a higher sensibility of adolescent breast tissues when compared with older women, with a higher incidence of mammary cancer in adolescent women when compared with adult ones (Mcgregor et al., 1977). The rat mammary gland is susceptible to distinct types of radiation, namely x-rays, neutron, gamma radiation and magnetic fields. They may be used alone or in combination with chemical carcinogens or hormones to induce mammary carcinogenesis in these animals (Russo and Russo, 1996; Thompson and Singh, 2000). Inversely to that observed in women, the incidence of neoplasms in Sprague-Dawley female rats exposed to X-rays was not different among the different aged groups. The mammary lesions identified were histologically classified as adenocarcinoma and fibroadenomas (Holtzman et al., 1982).

IMPLANTATION OF CANCER CELLS

Additionally to the techniques described above to the induction of mammary tumors, tumor cells may be directly implanted in rats for tumor development. Depending on the origin of the cancer cells, these models may be classified as xenograft or syngeneic. In the xenograft models, the tumor cells are derived from human mammary tumors and are implanted in immunocompromised animals (Kim, O'Hare, and Stein, 2004b). The rnu/rnu nude rats described above are the most frequently used for this kind of mammary cancer modelling (Marchesi, 2013; Yang et al., 2014). In the syngeneic models, the tumor cells had origin in laboratory animals genetically similar to those in which tumor cells will be implanted, avoiding the use of immunocompromised animals and the costs associated with their maintenance. The syngeneic models are useful to study the interaction between immune system and cancer development (Khanna and Hunter, 2005).

Both xenograft and syngeneic models may be orthotopic if the mammary tumor cells are implanted in the tumor site of origin (mammary fat pad), or heterotopic when the tumor cells are implanted in a different place (subcutaneously, intraperitoneally, intramuscular) (Sano and Myers, 2009; Vargo-Gogola and Rosen, 2007).

GENETICALLY-ENGINEERED MODELS

Advances in molecular biology and the ability to create genetically modified animals has completely changed our ability to understand the molecular mechanisms and cellular pathways underlying biologic processes and disease states, including cancer (Doyle et al., 2012; Vandamme, 2014). Genetically modified animals are organisms whose genetic material was changed by adding (transgenic), modifying (knockin) or removing (knockout) DNA sequences (Forabosco et al., 2013). Although the number of genetically engineered mice is significantly higher when compared with the number of genetically engineered rats, the genetically modified rats have become an excellent model to study aspects of molecular etiology of mammary cancer (European Commission, 2010; Smits et al., 2007). In the specific case of mammary carcinogenesis, the genetic engineered rats are excellent models to evaluate the role of *ras*, *neu*, *BRCA1* and *BRCA2* genes on the malignant progression of mammary tumors (Cotroneo et al., 2007a; Hoenerhoff et al., 2011; Mullins et al., 2002; Smits et al., 2007; Zan et al., 2003).

PATHOGENESIS OF RAT MAMMARY CANCER, SAMPLE COLLECTION AND HISTOLOGICAL EVALUATION

Mammary tumors in rats may arise from both parenchyma and stroma cells, and they may be composed of a single histological type or a combination of distinct patterns. A very complete and useful histological classification of chemically-induced rat mammary tumors taking into account their histogenesis and behavior was established by Russo and Russo (Russo and Russo, 2000). In general, the carcinogenesis initiates within 14 days after the administration of the carcinogen agent with an enlargement of terminal end bud. After initiation, the carcinogenesis may follow two distinct pathways: one in which benign lesions such as cysts, adenomas, alveolar hyperplasias and fibroade-nomas originate from alveolar buds, and another one in which terminal end bud originates malignant lesions (intraductal carcinomas) with distinct histological patterns, such as papillary, cribriform and comedo (Russo and Russo, 1998, 1996). Usually, chemically-induced mammary tumors are detected by palpation at 7th to 10th week after carcinogen administration and the number of tumors increases over time (Faustino-Rocha et al., 2016b, 2013d, 2015c; Russo et al., 1982). At sacrifice, animals were skinned/scalped and the skin was carefully evaluated under a light in order to detect small mammary tumors not previously detected by palpation (Faustino-Rocha et al., 2016b, 2013b, 2013d, 2016d; Lopes et al., 2014). The fixation of mammary tumors in buffered formalin followed by routine staining with hematoxylin and eosin (H&E) and observation in a light microscope is the most frequently used method to classify the chemically-induced rat mammary tumors.

MONITORING OF POTENTIAL THERAPEUTIC APPROACHES

The detection of breast alterations and the response of mammary tumors to pharmacological and non-pharmacological therapeutic strategies may be non-invasively evaluated through the use of different imaging techniques, namely mammography, ultrasonography, elasto-graphy and magnetic resonance imaging (MRI). Although several imaging modalities are available and they have improved the sensitivity of breast cancer detection and diagnosis, each of them has advantages and disadvantages, and should be selected taking into account some factors, such as patients' age and breast density (Andreea et al., 2011). Despite all the advances, it is important to take into account that no single imaging modality is able to identify and characterize breast abnor-malities, and a combined approach continues to be necessary. Indeed some of imaging techniques, such as mammography and sonoelas-tography facilitate lesion detection, while other techniques, namely ultrasound and magnetic resonance imaging (MRI) are more focused on the characterization of the lesions (Andreea et al., 2011).

Mammography is still considered the "gold" standard for the detection of breast alterations and it is the only widely used imaging technique for breast cancer screening. Mammography provides an adequate visualization of soft tissue abnormalities and the detection of subtle calcifications (Karellas and Vedantham, 2008). However, it has some disadvantages, namely radiation imposition, limited dynamic range, contrast characteristics, granularity, and difficulty in detecting very subtle lesions, especially in dense breasts (Nishikawa et al., 1987).

The importance of ultrasonography for breast screening has increased. Ultrasonography provides a panoramic high resolution image of entire breast and it is very useful to the detection of small cancers not detected in mammography, to the differentiation between benign and malignant solid breast lesions, and also to perform several eco-guided procedures, namely needle aspiration and core-needle biopsies (Corsetti et al., 2006; Gordon and Goldenberg, 1995; Kaplan, 2001; Kolb, Lichy, and Newhouse, 1998, 2002; Moss et al., 1999). Beyond the B mode, the use of Color Doppler, Power Doppler, Pulsed Doppler, B Flow and Contrast-enhanced ultrasound allowed the detection and characterization of blood flow in mammary tumors (Kubota et al., 2002). When compared with other imaging modalities, ultrasonography has some advantages, namely it is a real-time exam, it does not impose radiation, the apparatus are portable and less expansive when compared with other modalities, namely MRI, it may be performed in patients with metal implants and it is more adequate to claustrophobic people. However, it has also some disadvantages, namely it is operator-dependent and some problems related with posterior acoustic shadowing may occur (Kubota et al., 2002).

Computed tomography is also under investigation as another tool for breast cancer detection. It is not limited by tissue overlapping as occurs in mammography (Kalender et al., 2012; Lindfors et al., 2010).

Tissue elasticity imaging technology is based on the change of the hardness of a tissue when it is affected by disease, such as cancer. Krouskop and collaborators (Krouskop et al., 1998) verified that the elasticity of most malignant breast and prostate tissues was higher when compared with normal tissues. Sonoelastography has the ability to show the relative stiffness of lesions compared with the stiffness of surround-dding tissues. Stiff tissues appeared in blue or dark gray tints, while soft tissues appear red, green or bright shades of green (Hall, Zhu, and Spalding, 2003; Tardivon et al., 2007). Usually, malignant masses appear dark and have high contrast with background breast tissues, while benign masses appear lighter and have a lower contrast with background breast tissue (Hall et al., 2003). False positive and false negative results constitute the main disadvantage of sonoelastography (Giuseppetti et al., 2005; Tardivon et al., 2007).

MRI is the most sensitive imaging modality for breast cancer detection, detecting breast alterations that are occult on other imaging modalities (Harms, 1998; Morris, 2002). The use of contrast agents provide a three-dimensional anatomical image and information about tumor angiogenesis. MRI may be used in women with dense breasts, it does not impose radiation and may detect multi-focal breast lesions. It is also very useful in determining if the cancer has spread to the chest wall and the cancer recurrence after tumor surgical excision. MRI has been frequently employed in patients with high risk of breast cancer development (patients carrying mutation in breast cancer genes or with a family history of breast cancer development) and as adjuvant to mammography in order to clarify indeterminate findings (Orel and Schnall, 2001). The cost, requirement of contrast agents injection for functional imaging, no detection of calcifications, claustrophobia and long-time required for scanning are the main disadvantages of MRI (Kristoffersen Wiberg et al., 2002).

Magnetic resonance spectroscopy is a promising technique for breast cancer screening and assessment of response to treatment. This technique allow the quantitative characterization of total choline concentration that has been shown to be increased in malignant mammary tumors when compared with normal breast tissue (Baek et al., 2008), and reduced 24 hours after the first dose of chemotherapy (Joe and Sickles, 2014). Magnetic resonance spectroscopy is a non-invasive imaging modality, it does not require contrast injection and demonstrates improved sensitivity and specificity when used as an adjunct to breast MRI (Bartella et al., 2007; Meisamy et al., 2005).

Positron emission tomography (PET) is one of the most recent imaging modalities. A radioactive substance, usually radioactively labeled sugar, is injected into a vein of the patient and goes to places in the body where the cells are more active, namely cancerous tissues where cells are in constant division. These radioactive substances gives off small amount of radiation that is detected by a special PET scanner to form an image. A PET scan may be combined with computed tomography in order to provide both anatomical and functional view of the suspect cells. PET is also very useful in identifying involved axillary lymph nodes and distant metastasis (Carter, Allen, and Henson, 1989). Breast density is not a limitation for PET scan. PET has as the main disadvantage the costs, the lower sensitivity in detecting some breast tumors due to their small size, metabolic activity, histological type (benign mammary lesions will be negative on PET scan), microscopic tumor growth and proliferation (Avril et al., 2001). The lack of evidences demonstrating clear advantages when compared with other imaging modalities has limited the use of PET in the routine diagnosis of breast cancer.

Thermography is a non-invasive technique that creates a temperature map of the breasts based on infrared radiation. It hypothesized that due to higher metabolism and vascularization of tumors when compared with surrounding tissues, a higher temperature in a region would indicate the presence of a cancer. However, it has proved to have a low sensitivity and no advantages when compared with mammography for breast cancer diagnosis (Feig et al., 1977; Gershon-Cohen and haberman, 1964).

Digital breast thomosynthesis is a new imaging modality that acquires multiple low-dose mammographic projections through the breast. Its use improves the detection and characterization of invasive cancers when compared with mammography, probably due to the decrease of tissue superimposition in the breast (Friedewald et al., 2014; Kopans, 2014).

Diffusion-weighted imaging is a non-invasive technique that evaluate breast tissue microstructure on the basis of random thermal motion (Brownian motion) of water molecules. Since tumors have increased cellularity, they will exhibit "restricted" or reduced diffusion of water molecules. This technique will be useful not only for the detection of breast lesions, but also characterization and assessment of response to the treatment (Partridge and McDonald, 2013; Woodhams et al., 2011).

Diffuse optical imaging is a non-invasive mean to get metabolic information through the use of near-infrared tissue absorption of light to measure hemoglobin concentration. The use of this technology aims to characterize benign and malignant lesions and monitoring response to chemotherapy (Cerussi et al., 2002; Jiang et al., 2009; Kukreti et al., 2010; Pogue et al., 2001; Zhi et al., 2012).

As imaging techniques are improving, the imaging will continue having an important role to reduce breast cancer morbidity and mortality.

Since the rat model of mammary cancer, specially the model of chemically-induced mammary cancer by DMBA and MNU, has been frequently used to evaluate potential therapeutic strategies for breast cancer, the imaging modalities previously presented for woman breast screening are also applicable for the evaluation of rat mammary tumors. Our research team has used the thermography and ultrasonography to monitor not only the growth but also the vascularization of mammary tumors MNU-induced in female Sprague-Dawley rats. The use of ultrasound allowed to determine the shape of growth and volume of mammary tumors by the application of different formulas and detect a higher vascularization of mammary tumors from exercised groups when compared with the sedentary ones (Faustino-Rocha et al., 2013a, 2016b, 2016, 2013b, 2013d, 2016e).

EVIDENCING DRUGS' EFFECTS IN RAT MAMMARY CANCER: POSITIVE ANTICANCER EFFECT OF ANTIHISTAMINES

Antihistamines are inverse agonists that inhibit the linkage of histamine to its receptors (Rimmer and Church, 1990; Van Schoor, 2012). They have a molecular structure similar to the histamine with which they compete (Sudo, Monsma, and Katzenellenbogen, 1983). The antihistamines have been routinely used for the treatment of several clinical conditions, namely motion sickness, insomnia, vertigo, and allergic diseases (contact dermatitis, atopic dermatitis, dermatoses, rhinitis, allergic conjunctivitis, mild transfusion reactions, urticarial and hypersensitivity reactions to drugs) (Carson, Lee, and Thakurta, 2010; Nadalin, Cotterchio, and Kreiger, 2003; Pharmacists, 2014; Van Schoor, 2012). The effects of antihistamines vary among patients (Van Schoor, 2012). Despite their safety and appropriate use is not fully clarified, the antihistamines are frequently used in children and adults (Criado et al., 2010; Del Cuvillo et al., 2006; Van Schoor, 2012). The knowledge of antihistamines pharmacokinetics and pharmacodynamics is essential to their correct use (Criado et al., 2010; Del Cuvillo et al., 2006; Van Schoor, 2012). Their dose should be adjusted in patients with renal of hepatic diseases (Del Cuvillo et al., 2006).

Antihistamines may be administered orally or topically applied (McCarthy et al., 2011; Nadalin et al., 2003). Antihistamines have a good absorption and reach the plasma concentration within three hours after oral administration (Simons, 2004). They are mainly biotransformed in the liver by the cytochrome enzyme system (CYP) (Bartra et al., 2006; Chen et al., 2003; Hamelin et al., 1998; Hey et al., 1999; Simons et al., 2002).

Table 1. *In vivo* studies using different rodent models of mammary cancer to assess the efficacy of several therapeutic strategies

Model/gender and animal strain	Drugs and/or compounds evaluated	Dose/treatment	Therapeutic effects	Ref.
DMBA-induced mammary tumors				
♀ Sprague-Dawley rats	Allyl isothiocyanate	Oral gavage (10, 20 and 40 mg/kg/day, 3 times/week for 16 weeks)	Exerted chemopreventive effects at a doses of 20 and 40 mg/kg	(Rajakumar, Pugalendhi, and Thilagavathi, 2015)
	Apple extract	Oral gavage (0, 3.3, 10 and 20 mg/kg b.w. for 26 weeks)	Inhibited mammary carcinogenesis	(Liu et al., 2009)
	Arthrospira platensis (Spirulina)	Diet supplementation (1% for 12 months)	Inhibited mammary carcinogenesis	(Ouhtit et al., 2014)
	Asiaticoside	2 weeks before and 8 weeks after DMBA administration or 5 weeks after DMBA administration (i.p., 200 µg/animal, 2 times/week)	Inhibited mammary carcinogenesis	(Al-Saeedi, 2014)
	Atrazine	Diet supplementation (5, 50 and 500 ppm for 34 weeks)	Enhanced tumor growth	(Ueda et al., 2005)
	Azadirachta indica	Diet supplementation (10-12.5 % for 2 weeks)	Inhibited mammary carcinogenesis	(Tepsuwan, Kupradinun, and Kusamran, 2002)
	Bamboo extract	Diet supplementation (0.5 % for 13 weeks)	Inhibited mammary carcinogenesis	(Lin et al., 2008)
	5,6-benzoflavone (5,6-BF) Indole-3-carbinol (I3C) Diindolylmethane (DIM)	5,6-BF (diet supplementation, 165, 550 and 1650 ppm, for 14 days) I3C (oral gavage, 180 mg/kg b.w. for 14 days) DIM (oral gavage, 20 and 180 mg/kg b.w. for 14 days)	5,6-BF and I3C were highly effective on mammary tumors inhibition; DIM had minimal effects	(Lubet et al., 2011)

Model/gender and animal strain	Drugs and/or compounds evaluated	Dose/treatment	Therapeutic effects	Ref.
	Bisphenol A	Oral administration (0, 25 and 250 µg/kg b.w. for 130 days)	Increased number of tumors and decreased latency period	(Lamartiniere et al., 2011)
	Broccoli sprouts	Oral gavage (1 mL for 5 days)	Inhibited mammary carcinogenesis	(Fahey, Zhang, and Talalay, 1997)
	4'-bromoflavone	Diet supplementation (2000 and 4000 mg/kg diet for 2 weeks)	Inhibited mammary carcinogenesis	(Song et al., 1999)
	Buckwheat protein extract	Diet supplementation (38.1% for 61 days)	Inhibited mammary carcinogenesis	(Kayashita et al., 1999)
	Calotropis procera protein Cyclophosphamide	Calotropis procera (0.2 mg/kg/b.w., oral administration for 20 days) Cyclophosphamide (0.2 mg/kg/b.w., i.p. for 20 days) Calotropis procera + Cyclophosphamide (0.1 mg/kg/b.w., oral for 20 days+ 0.1 mg/kg/b.w., i.p. for 20 days)	Inhibited mammary carcinogenesis	(Samy et al., 2012)
	3-carbamoyl-2,2,5,5-tetramethylpyrroline-1-oxyl (Pirolin) 2-(3,4-dihydroxyphenyl)-3,5,7-trihydroxychromen-4-one (Quercetin)	i.p. administration (10 mg/kg, for 14 days)	Acted as cytoprotector and inhibited tumor growth	(Tabaczar et al., 2015)
	Celecoxib	Diet supplementation (500 ppm and 1500 ppm or 15 weeks)	Inhibited mammary carcinogenesis	(Jang et al., 2002)
	Chorionic gonadotropin	i.p. administration (100 IU, for 5, 10, 20 and 40 days)	Induced cell death	(Srivastava, Russo, and Russo, 1997)

Table 1. (Continued)

Model/gender and animal strain	Drugs and/or compounds evaluated	Dose/treatment	Therapeutic effects	Ref.
	Cimicifuga racemosa extract.	Oral gavage (0.714, 7.14 and 71.4 mg/kg b.w. for 6 weeks)	No significant effects were observed	(Freudenstein, Dasenbrock, and Nisslein, 2002)
	Cisplatin Nordihydroguaiaretic acid (NDGA)	After tumor development: NDGA (10 mg/kg, i.p., for 5 days) followed by cisplatin (7.5 mg/kg, i.p., single dose)	Reduced kidney toxicity and improved anti-breast cancer activity	(Mundhe et al., 2015)
	Cloudy apple juice	Oral gavage (10 mL/kg/b.w., for 28 days before DMBA administration)	Decreased blood levels of biochemical liver and kidney markers	(Szaefer et al., 2014)
	Conjugated linoleic acids	Diet supplementation (1.0 or 2.0%) before, after, or before and after the DMBA administration for 23 weeks	Inhibited mammary carcinogenesis	(Bialek, Tokarz, and Zagrodzki, 2013)
	Copper Resveratrol	Copper (42.6 mg/kg diet, oral gavage for 100 days) Copper + Resveratrol (42.6 mg/kg diet + 0.2 mg/kg b.w., oral gavage for 100 days)	Promoted tumor growth	(Skrajnowska et al., 2013)
	Cow's milk	Oral administration (for 21 days after birth)	Reduced risk of mammary tumors development	(Nielsen et al., 2011)
	Curcumin Dibenzoylmethane	Curcumin (diet supplementation, 0.2 and 1% for 14 days) Dibenzoylmethane (diet supplementation, 0.5 and 1% for 14 days)	Inhibited mammary carcinogenesis	(Singletary et al., 1998)

Model/gender and animal strain	Drugs and/or compounds evaluated	Dose/treatment	Therapeutic effects	Ref.
	Diethylstilbestrol	s.c. administration (1 µg/animal, 0-14, 0-5 or 6-14 days after birth)	The administration from 0 to 14 days after birth inhibited mammary carcinogenesis	(Yoshikawa et al., 2008)
	2,2'-diphenyl-3,3'-diindolylmethane	Oral gavage (5 mg/kg b.w., each two days for 21 days)	Inhibited mammary carcinogenesis	(Bhowmik et al., 2013)
	Eicosapentaenoic acid (EPA) Docosahexaenoic acid (DHA)	Oral gavage (0.5 mL for 20 weeks)	Inhibited mammary carcinogenesis	(Noguchi et al., 1997)
	Endostatin	s.c. administration (20 mg/kg for 28 dyas)	Inhibited mammary carcinogenesis	(Perletti et al., 2000)
	Enterolactone	Oral gavage (1 and 10 mg/kg b.w. for 50 days)	Inhibited mammary carcinogenesis	(Saarinen et al., 2002)
	9α-Fluoromedroxyprogesterone Acetate (FMPA)	Oral administration (30 and 120 mg/Kg for 3 weeks)	Inhibited mammary carcinogenesis	(Murata et al., 2006)
	Folic acid	Diet supplementation (5, 8 and 10 mg/kg diet, after tumor development, for 12 weeks)	Promoted the mammary tumors progression	(Deghan Manshadi et al., 2014)
	Ganoderma lucidum	Oral gavage (500 mg/kg b.w. for 16 weeks)	Inhibited mammary carcinogenesis	(Deepalakshmi and Mirunalini, 2013)
	Genistein Daidzein	Genistein + Daidzein (oral administration, 20 mg + 20 mg/kg, for 16 weeks)	Protected the structural integrity of cell surface and membranes	(Pugalendhi et al., 2011)
	Grape seed extract	Grape seed extract (diet supplementation, 1.25 and 5 % for 135 days)	No protective effects were observed	(Kim et al., 2004a)

Table 1. (Continued)

Model/gender and animal strain	Drugs and/or compounds evaluated	Dose/treatment	Therapeutic effects	Ref.
	Green tea polyphenol Black tea polyphenol	Green and black tea polyphenols (oral administration, 0.1% for 28 weeks)	Both polyphenols inhibited mammary carcinogenesis	(Roy et al., 2011)
	Iodine Potassium iodide	Iodine (0.05%, 0.07%) Potassium iodide (0.05%, 0.1%) Iodine + Potassium iodide (0.05% + 0.05%; 0.05% + 0.1%)	The combination of Iodine + Potassium iodide (0.05% + 0.05%) exerted antineoplastic effects	(Soriano et al., 2011)
	β-ionone	Diet supplementation (9, 18 and 36 mmol/kg for 24 weeks)	Inhibited mammary carcinogenesis	(Liu et al., 2008)
	Iron	s.c. administration (50 μmol/kg, twice a week, for 53 weeks)	Promoted mammary carcinogenesis	(Diwan, Kasprzak, and Anderson, 1997)
	Isoflavone	Diet supplementation (100, 500 or 1000 mg/kg diet, for 24 weeks)	Inhibited mammary carcinogenesis (decreased tumor incidence, mean number of tumors per animals and increase tumor latency)	(Ma et al., 2014)
	Kalpaamruthaa	Oral gavage (100, 200, 300, 400 and 500 mg/kg b.w., for 14 days)	Inhibited mammary carcinogenesis	(Veena, Shanthi, and Sachdanandam, 2006)
	Low-fat milk (1%) Artificial milk Estrone sulfate solution (0.1 μg/mL)	Oral administration for 20 weeks	Low-fat milk and estrone promoted mammary carcinogenesis	(Qin et al., 2004)

Model/gender and animal strain	Drugs and/or compounds evaluated	Dose/treatment	Therapeutic effects	Ref.
	Magnetic fields	50-Hz twice a week, for 18 weeks	Promoted mammary carcinogenesis	(Fedrowitz, Kamino, and Löscher, 2004)
	Medroxyprogesterone acetate (MPA) Norgestrel (N-EL) Norethindrone (N-ONE) Megestrol acetate (MGA)	MPA, N-EL, N-ONE, MGA (s.c. implant, 60 days)	N-EL inhibited tumor growth, MGA did not change tumor growth	(Benakanakere et al., 2010)
	Melatonin	Drinking water (25 µg/mL for 9 weeks)	Inhibited mammary carcinogenesis	(Cos et al., 2006)
		Oral gavage (10 mg/kg for 15 days and 6 months)	Inhibited mammary carcinogenesis	(Clarke, 1996)
	Methyl-amoorain (methyl-25-hydroxy-3-oxoo-lean-12-en-28-oate)	Oral gavage (0.8, 1.2 and 1.6 mg/kg b.w., 3 times/week for 18 weeks)	Promoted mammary carcinogenesis	(Mandal, Bhatia, and Bishayee, 2013)
	Methylseleninic acid	Diet supplementation (2 ppm for 22 weeks)	Antineoplastic effects	(Ip et al., 2000)
	Milk Estrone sulfate	Milk (oral administration for 20 weeks) Estrone sulfate (100 ng/ml for 20 weeks)	Both promoted mammary carcinogenesis	(Ma et al., 2007)
	Morin (3,5,7,2,'4'-pentahydroxyflavone)	Oral administration (50 mg/kg b.w., thrice a week, for 15 weeks)	Beneficial effects as chemopreventive agent	(Nandhakumar, Salini, and Niranjali Devaraj, 2012)
	N-(4-hydroxyphenyl)retinamide-C- glucoronide (4-HPRCG)	Diet supplementation (2 mmol/kg diet for 28 days)	Inhibited mammary carcinogenesis	(Alshafie et al., 2005)

Table 1. (Continued)

Model/gender and animal strain	Drugs and/or compounds evaluated	Dose/treatment	Therapeutic effects	Ref.
	N-acetyl-L-cysteine (NAC) Anethole trithione Phenethylisothiocyanate (PEITC) Miconazole	NAC (diet supplementation, 4000 and 8000 ppm for 100 days) Anethole trithione (diet supplementation, 200 and 400 ppm for 100 days) PEITC (diet supplementation, 600 and 1200 ppm for 100 days) Miconazole (diet supplementation, 1000 and 2000 ppm for 100 days)	Anethole trithione and miconazole inhibited mammary carcinogenesis PEITC promoted mammary carcinogenesis No effects were observed for NAC	(Lubet et al., 1997)
	Operculina turpethum	Oral gavage (100 mg/kg b.w. for 45 days)	Decreased tumor weight	(Anbuselvam, Vijayavel, and Balasubramanian, 2007)
	Paclitaxel *Eruca sativa* seeds	Pacliatxel encapsulated liposome (intravenous, 20 mg/kg/week) and *Eruca sativa* seeds extract (oral, 500 mg/kg/week) for 4 weeks	Reduced inflammation and cell proliferation	(Shaban et al., 2016)
	Photodynamic therapy	Photodithazine (8 mg/kg, i.p.) and 100 J/cm of light at a fluence rate of 100 mW/cm	Upregulation of apoptotic genes and downregulation of anti-apoptotic genes	(Silva et al., 2014)
	Plasticizer benzyl butyl phthalate	i.p. administration or oral gavage (100 and 500 mg/kg for 7 days)	Inhibited mammary carcinogenesis	(Singletary, MacDonald, and Wallig, 1997)
	Pleurotus ostreatus	Oral gavage (150, 300 and 600 mg/kg b.w. for 16 weeks)	Inhibited mammary carcinogenesis	(Krishnamoorthy and Sankaran, 2013)
	Pomegranate	Oral gavage (0.2 g/kg, 1.0 g/kg or 5.0 g/kg, 3 times/week) for 18 weeks (2 weeks before and 16 weeks after DMBA administration)	Chemopreventive effects	(Bishayee et al., 2016)

Model/gender and animal strain	Drugs and/or compounds evaluated	Dose/treatment	Therapeutic effects	Ref.
	Pomegranate	Diet supplementation (0.2, 1.0 and 5.0 mg/kg) for 18 weeks (2 weeks before and 16 weeks after DMBA administration)	Inhibited mammary carcinogenesis. Inhibited proliferation and promoted the apoptosis	(Mandal and Bishayee, 2015a)
	RS100642	Intravenous (0.25 mg/kg b.w., once a week for 4 weeks)	Improved survival	(Batcioglu et al., 2012)
	RU486 CDB-4124	s.c. administration, 10 mg/kg b.w., for 28 days	Inhibited mammary carcinogenesis	(Wiehle, Christov, and Mehta, 2007)
	Selenium-enriched Japanese radish sprout	Diet supplementation (8.8 ppm for 13 or 28 weeks)	Inhibited mammary carcinogenesis	(Yamanoshita et al., 2007)
	Shemamruthaa	Oral gavage (400 mg/kg b.w., for 14 days)	Inhibited mammary carcinogenesis	(Purushothaman, Nandhakumar, and Sachdanandam, 2013)
	Simvastatin	Oral gavage (20 and 40 mg/kg, for 14 days)	Reduced tumor growth	(Rennó et al., 2015)
	Soy milk	Oral administration for 20 weeks	Promoted mammary carcinogenesis	(Qin et al., 2007)
	Taurine	Drinking water (3%) for 16 weeks	Inhibited mammary carcinogenesis	(He, Li, and Guo, 2016)
	Taurine	Oral gavage (100 mg/kg) for 5 weeks	Efficient as chemotherapeutic agent	(Vanitha et al., 2015)
	Tamoxifen Quercetin	Tamoxifen (3 mg/kg, oral gavage for 3 days) Tamoxifen + Quercetin (3 mg/kg + 6 mg/kg, oral gavage for 3 days)	Inhibited mammary tumors angiogenesis	(Jain, Thanki, and Jain, 2013)
	Tangeretin (4',5,6,7,8-pentamethoxyflavone)	Oral administration (50 ng/kg, pre-treatment for 30 days or post-treated for 30 days)	Efficient as chemotherapeutic agent	(Periyasamy et al., 2015)

Table 1. (Continued)

Model/gender and animal strain	Drugs and/or compounds evaluated	Dose/treatment	Therapeutic effects	Ref.
	Trianthema portulacastrum	Diet supplementation (50, 100 and 200 mg/kg) for 18 weeks (2 weeks before and 16 weeks after DMBA administration)	Reduced inflammation and suppressed mammary carcinogenesis	(Mandal and Bishayee, 2015b)
	Trianthema portulacastrum	Diet supplementation (50, 100 and 200 mg/kg body weight, for 18 weeks)	Inhibited mammary carcinogenesis	(Mandal and Bishayee, 2015c)
	Tualang Honey	Oral administration (0.2, 1.0 or 2.0 g/kg b.w. for 150 days)	Inhibited mammary carcinogenesis	(Kadir et al., 2013)
	Vanadium Fish oil	Vanadium (drinking water, 0.5 ppm for 6 weeks) Fish oil (oral gavage, 0.5 mL/day for 6 weeks) Vanadium + Fish oil (drinking water, 0.5 ppm + oral gavage, 0.5 mL/day, for 6 weeks)	Vanadium and fish oil alone were effective on mammary tumors inhibition, the combination exhibited higher effectiveness	(Manna et al., 2011)
	Wheat bran fiber	Diet supplementation (5, 9.6 and 17.6% for 13 weeks)	Inhibited mammary carcinogenesis	(Zile, Welsch, and Welsch, 1998)
	Zearalenone Genistein	Zearalenone and Genistein (s.c. administration, 20 µg for 5 days)	Inhibited mammary carcinogenesis	(Hilakivi-Clarke et al., 1999)
	Zinc Resveratrol Genistein	Zinc (231 mg/kg diet, oral gavage for 40 days) Zinc + Resveratrol (231 mg/kg diet + 0.2 mg/kg b.w., oral gavage for 40 days) Zinc + Genistein (231 mg/kg diet + 0.2 mg/kg b.w., oral gavage for 40 days)	The combination of Zinc + Resveratrol increased tumor number	(Bobrowska-Korczak, Skrajnowska, and Tokarz, 2012)
♀ Wistar-Furth rats	Limonene	Diet supplementation (0, 2.5, 5.0, 7.5 and 10% for 11 weeks)	Regression of mammary tumors above 7.5%	(Haag, Lindstrom, and Gould, 1992)

Model/gender and animal strain	Drugs and/or compounds evaluated	Dose/treatment	Therapeutic effects	Ref.
♀ Wistar rats	BAY 12-9566N	Diet supplementation (240 mg/kg/day for 52-64 weeks)	Inhibited mammary carcinogenesis	(Iatropoulos et al., 2008)
	Celecoxib n-3 polyunsaturated fatty acids (PUFA)	Oral administration of celecoxib (20 mg/kg) in combination with PUFA (0.5 mL) for 7 days	Chemopreventive effect in mammary carcinogenesis	(Negi, Bhatnagar, and Agnihotri, 2016)
	Crateva adansonii DC	Oral gavage (75 and 300 mg/kg b.w., once by day, for 13 weeks)	Inhibited mammary carcinogenesis, mainly at a dose of 75 mg/kg b.w.	(Zingue et al., 2016)
	Lycopene Genistein	Lycopene (20 mg/kg b.w., oral gavage, thrice a week for 20 weeks) Genistein (2 mg/kg b.w., oral gavage, thrice a week for 20 weeks) Lycopene + Genistein (20 mg/kg b.w. + 2 mg/kg b.w., oral gavage, thrice a week)	Inhibition of mammary carcinogenesis	(Sahin et al., 2011)
	Melissa officinalis	Oral gavage (100 mg/kg b.w. for 4 weeks after tumors development)	Inhibited tumor growth	(Saraydin et al., 2012)
	Organoselenium compounds	i.p. administration (25 µmol/kg, each two days for four weeks)	Inhibited mammary carcinogenesis	(Ozdemir et al., 2006)
	Vincristine Myricetin	Vincristine (i.p., 500µg/kg, 1 administration/week, for 4 weeks) Myricetin (oral, 50, 100 and 200 mg/kg, every day, 16 weeks)	Each drugs, independently, inhibited mammary carcinogenesis	(Jayakumar et al., 2014)
♀ Zucker rats	Casein Soy protein + isoflavones	Casein (diet supplementation, 200 g/kg/diet for 7 weeks) Soy protein + isoflavones (diet supplementation, 202 g/kg/diet + 3.24 mg/g protein for 7 weeks)	Casein inhibited mammary carcinogenesis	(Hakkak et al., 2011)

Table 1. (Continued)

Model/gender and animal strain	Drugs and/or compounds evaluated	Dose/treatment	Therapeutic effects	Ref.
♀ Donryu rats	Dehydroepiandrosterone	Diet supplementation (0.6% for 15 weeks)	Inhibited mammary carcinogenesis	(Hakkak et al., 2010)
♀ Donryu rats	Isoflavone aglycones	Diet supplementation (0.2% for 2 weeks, 4 weeks or 40 weeks)	Promoted mammary carcinogenesis	(Kakehashi et al., 2012)
♀ Holtzman rats	*Piper aduncum*	Oral administration/capsules (50, 150 and 300 mg/kg/body weight)	Decreased mammary carcinogenesis and lymph node metastasis	(Arroyo-Acevedo et al., 2015)
♀ albino rats	*Nigella sativa* Thymoquinone	Oral gavage (1, 5 and 10 mg/kg, 3 times/week, for 4 months)	Inhibited mammary carcinogenesis	(Linjawi et al., 2015)
♀ rats	Apigenin	Diet supplementation (0.02, 0.1 and 0.5 % for	Promoted mammary carcinogenesis	(Mafuvadze et al., 2013)
MNU-induced mammary tumors				
♀ Lewis rat	Anastrazole Docetaxel HE3235 + Docetaxel 17α-ethynyl-5α-androstane-3α, 17β-diol (HE3235) Tamoxifen	i.p. administration (2.5 mg/day for 4 weeks) i.p. administration (1.5 mg, once a week for 4 weeks) i.p. administration (6.6 mg/day + 1.5 mg, once a week for 4 weeks) i.p. administration (4-6.6 mg/day for 4 weeks) s.c. administration (0.25 mg, once a week for 4 weeks)	All compounds decreased incidence and number of mammary tumors; the high dose of HE3235 in combination with docetaxel was the most efficient treatment	(Ahelm et al., 2011)
♀ Wistar rat	Carboxy ethyl germanium sesquioxide (Ge-132)	i.p. administration (1500 mg/kg/day for 34 weeks)	Reduced tumors' growth	(Vinodhini and Sudha, 2013)

Model/gender and animal strain	Drugs and/or compounds evaluated	Dose/treatment	Therapeutic effects	Ref.
	Green tea extract	Diet supplementation (30 mg for 9 weeks)	Tumor multiplicity was lower in animals that received green tea extract	(Kale et al., 2010)
♀ Ludwig/Wistar/ Olac rat	Pamidronate	s.c. (0.4 mg/kg/b.w. for 4 weeks)	Reduced tumor volume	(Colston et al., 2003)
♀ rat	Potato (*Solanum tuberosum* L.)	Diet supplementation (5-50% for 5 weeks)	Reduced cancer incidence	(Thompson et al., 2009)
♀ Sprague-Dawley rat	Amphetamine-regulated transcript peptide (CART)	Intracerebroventricular (1 μg/rat/day, for 5 days) Intracerebroventricular (5 μl (1:500)/rat/day, for 5 days)	Reversed cancer cachexia	(Nakhate et al., 2010)
	Anastrazole	Diet supplementation (0.05-0.5 mg/kg for 15 weeks)	High concentration reduced mammary tumors incidence and number of tumors *per* animal	(Kubatka et al., 2008a)
	Carboplatin Methrotrexate Paclitaxel	Intraductal (6 mg/rat, single administration) Intraductal (4-10 mg/rat, single administration) Intraductal (60 mg/rat, single administration)	Carboplatin was the most efficient agent in the inhibition of mammary carcinogenesis	(Stearns et al., 2011)
	Celecoxib	Diet supplementation (1500 ppm, 7-24 weeks)	Suppressed mammary carcinogenesis	(Badawi et al., 2004; Lu et al., 1997)

Table 1. (Continued)

Model/gender and animal strain	Drugs and/or compounds evaluated	Dose/treatment	Therapeutic effects	Ref.
	13-cis retinoic acid (13cRA) CpG oligodeoxynucleotides (CpG-ODN) 13-cis retinoic acid (13cRA) + CpG oligodeoxynucleotides	Intragastric (1 mg/kg, 3 times a week for 15 weeks) Intraductal (CpG-ODN motifs, 2 administrations) Intragastric (1 mg/kg, 3 times a week, 15 weeks + 2 administrations)	CpG-ODN reduced the number of mammary tumors	(Liska et al., 2003)
	Curcumin	Intraductal (168 µg encapsulated drug/treatment, 2-3 administrations) Oral gavage (200 mg/kg/b.w., 2-3 adminsitrations)	Reduced the incidence of mammary tumors	(Chun et al., 2012)
	Doxorubicin (DOX) DOX + Iodine (I2)	i.p. administration (4-16 mg/kg, 1 day) i.p. administration (4-16 mg/kg, 1 day) +Drinking water (0.05% for 7 days)	I_2 may be used as adjuvant of doxorubicin in cancer therapy	(Alfaro et al., 2013)
	Exemestane	Diet supplementation (1-10 mg/kg for 13 weeks)	Administration in premenopausal animals induced mammary carcinogenesis	(Kubatka et al., 2008a)
	Fluorouracil	i.v. (12 mg/rat, 4 administrations) Intraductal (12 mg/rat, 4 administrations)	Intraductal administration inhibited mammary carcinogenesis	(Stearns et al., 2011)
	Flurbiprofen	Diet supplementation (31.25-62.5 mg/kg for 26 weeks)	Inhibited mammary carcinogenesis	(Mccormick and Moon, 1983)
	Garlic powder S-all1cysteine (SAC) Diallyl disulfide (DADS)	Diet supplementation (20 g/kg for 27 weeks) Diet supplementation (57 nmol/kg for 27 weeks) Diet supplementation (57 µmol/kg for 27 weeks)	Inhibited mammary carcinogenesis	(Schaffer et al., 1996)
	Genistein	s.c. administration (12.5 mg/day for 3 days)	Promoted mammary carcinogenesis	(Yang et al., 2000)

Model/gender and animal strain	Drugs and/or compounds evaluated	Dose/treatment	Therapeutic effects	Ref.
	High fat, low fiber diet + phytic acid	Diet supplementation (2% phytic acid from 9-30 weeks)	Phytic acid contributed to the reduction of mammary tumors incidence	(Shivapurkar et al., 1996)
	1α-Hydroxyvitamin D_5	Diet supplementation (25-50 µg/kg for 18 weeks)	Inhibited mammary carcinogenesis	(Mehta et al., 2000)
	Keoxifene	s.c. administration (20-500 µg for 13 weeks)	Inhibited mammary carcinogenesis	(Gottardis and Jordan, 1987)
	Ketotifen	Drinking water (1 mg/kg for 18 weeks)	Inhibited mammary tumor development.	(Faustino-Rocha et al., 2014)
	Lapatinib	Oral gavage (25-75 mg/kg/b.w. for 21 weeks)	High dose inhibited mammary carcinogenesis	(Li et al., 2011)
	Letrozole	Diet supplementation (1-10 mg/kg for 18 weeks)	Inhibited mammary carcinogenesis	(Kubatka et al., 2008b)
	Lysine, arginine, proline, ascorbic acid and green tea extract	Diet supplementation (0.5% for 24 weeks)	Inhibited mammary carcinogenesis	(Roomi et al., 2005)
	Mango (*Mangifera indica* L.)	Drinking water (0.02-0.06 g/ml for 2 or 23 weeks)	Did not inhibit mammary carcinogenesis	(Garcia-Solis, Yahia, and Aceves, 2008)
	2-methoxyestradiol	s.c. administration (1-5 mg/kg/day for 4 weeks)	Did not inhibit mammary carcinogenesis	(Lippert et al., 2003)
	Paclitaxel	Intraductal (10-25 mg/kg, for 8 weeks) i.p. administration (25 mg/kg for 8 weeks)	Local administration may reduce mammary carcinogenesis	(Okugawa et al., 2005)

Table 1. (Continued)

Model/gender and animal strain	Drugs and/or compounds evaluated	Dose/treatment	Therapeutic effects	Ref.
	Potassium iodide (KI) Iodine (I₂) Thyroxine (T4)	KI and I₂ (0.05% in drinking water from 3-18 weeks) T4 (3 µg/ml in drinking water from 3-18 weeks)	Long-term I₂ treatment inhibited mammary carcinogenesis	(Garcia-Solis et al., 2005)
	Raloxifene Keoxifen (9-cis-retinoic acid + raloxifene)	Diet supplementation (20-60 mg/kg for 19 weeks) Diet supplementation (60 mg/kg + 20-60 mg/kg)	Inhibited mammary carcinogenesis	(Anzano et al., 1996)
	Resveratrol (trans-3,4',5-trihydroxystilbene)	s.c. administration (10-100 mg/kg/day for 5 days)	Prepubertal treatment promoted mammary carcinogenesis	(Sato et al., 2003)
	Retinoid 9cUAB30 + Tamoxifen	Diet supplementation (150 mg/kg + 0.4 mg/kg for 21 weeks)	Combination of both agents inhibited mammary carcinogenesis	(Grubbs et al., 2003)
	Tamoxifen	s.c. administration (6.25-500 µg for 8 weeks) i.p. administration (1 mg/kg for 8 weeks)	Inhibited mammary carcinogenesis	(Gottardis and Jordan, 1987; Martin et al., 1996)
	Targretin	Oral gavage (6.7-60 mg/kg/day for 17 weeks) Diet supplementation (92-275 mg/kg for 17 weeks)	Inhibited mammary carcinogenesis	(Lubet et al., 2005)
Irradiation-induced mammary tumors				
♀ Sprague-Dawley rats	Estriol Estradiol	s.c. administration (638 µ/month for 6 months)	Delayed mammary tumors development	(Lemon et al., 1989)

Model/gender and animal strain	Drugs and/or compounds evaluated	Dose/treatment	Therapeutic effects	Ref.
Hormone-induced mammary tumors				
♀ August Copenhagen Irish rats (estrogen-induced)	Resveratrol	s.c. pellet (50 mg for 8 months)	Inhibited mammary carcinogenesis	(Singh et al., 2014)
♀ ACI rats (estrogen-induced)	Dietary restriction	Diet restriction (40% less energy)	Inhibited mammary carcinogenesis	(Harvell et al., 2002)
	Tocopherol	Diet supplementation (0.3% for 14 days)	Inhibited mammary carcinogenesis	(Das Gupta et al., 2015)
	Vitamin C Butylated hydroxyanisole	Vitamin C (1% drinking water, for 15 days) Butylated hydroxyanisole (diet supplementation, 0.7% for 120 days)	Vitamin C and Butylated hydroxyanisole inhibited mammary carcinogenesis	(Singh, Bhat, and Bhat, 2012)
♀ ACI rats (17β-estradiol and testosterone)	Phanobarbital	Drinking water (0.05% for 6, 12 and 28 weeks)	Inhibited mammary carcinogenesis	(Mesia-Vela et al., 2006)
	Tamoxifen	Subpanicular implant (40 mg for 6 months)	Inhibited mammary carcinogenesis	(Li et al., 2002)
		Implantation (40 mg for 4 and 7 months)	Inhibited mammary carcinogenesis	(Montano et al., 2007)
♀ ACLCOP-Ept2 rats (17β-estradiol and testosterone)	Tamoxifen	s.c. implantation (for 5 months)	Inhibited mammary carcinogenesis	(Ruhlen et al., 2009)
Xenograft model of mammary cancer				
Tumor cells inoculation (T47-D and BT474) in	Medroxyprogesterone acetate 3-(5'-hydroxymethyl-2'-furyl)-1-benzylindazole (YC-1)	Medroxyprogesterone acetate pellets + YC-1 (10 mg/60-day release + YC (i.p., 600 µg)	Inhibited mammary carcinogenesis	(Carroll et al., 2013)

Table 1. (Continued)

Model/gender and animal strain	Drugs and/or compounds evaluated	Dose/treatment	Therapeutic effects	Ref.
Sprague-Dawley rats 24-48 hours after the implantation of a pellet of 17β-estradiol (1.7 mg, 60 days timed release)	Medroxyprogesterone acetate 3-(5'-hydroxymethyl-2'-furyl)-1-benzylindazole (YC-1)	Medroxyprogesterone acetate pellets + YC-1 (10 mg/60-day release + YC (i.p., 600 μg)	Inhibited mammary carcinogenesis	(Carroll et al., 2013)
Syngeneic model of mammary cancer				
Tumor cells inoculation (MT-450) in Wistar Furth rats	VEGF-C or VEGFR-3 antibodies	Intradermal injection (5 times/week for 4 weeks)	Inhibited lung metastasis	(Quagliata et al., 2014)
	Delphinidin	Oral administration (1.18×10^{-5} mol for 28 days)	Promoted tumor growth and metastasis	(Thiele et al., 2013)
Tumor cells injection (i.v., MADB106) in Fischer 344 rats	CpG oligodeoxynucleotides (CpG-C ODN)	i.p. injection (330 μg/kg)	Reduced lung retention	(Goldfarb et al., 2009)
	Interleukin-12 (4×1.5 μg/kg)	s.c. injection (0.5μg/rat/0.5 ml injection for 8 days)	Reduced lung retention	(Avraham et al., 2010)
Tumor cells inoculation (Mat B III) in Fischer 344 rats	Antiangiogenic Urokinase-derived Peptide (Å6) Tamoxifen	Å6 (i.p., 75 mg/kg/day for 17 days) TAM (i.p., 3 mg/kg/day for 17 days) Å6 + Tamoxifen (i.p., 75 mg/kg/day + 3 mg/kg/day for 17 days)	The Å6 enhanced the antitumor effects of Tamoxifen	(Guo et al., 2002)
	Rat umbilical cord matrix stem cells (rUCMS)	Intratumoral or i.v. injection (2 administartions)	Attenuated mammary cancer growth	(Ganta et al., 2009)

Model/gender and animal strain	Drugs and/or compounds evaluated	Dose/treatment	Therapeutic effects	Ref.
	Tamoxifen 4-iodo benzo[b]thiophene-2-carboxamidine (B-428)	Tamoxifen (i.p. administration, 3 mg/kg for 2 weeks) B-428 (i.p. administration, 0.005 ml/h for 2 weeks) Tamoxifen + B-428 (i.p. administration, 3 mg/kg + 0.005 ml/h for 2 weeks)	The combination inhibited mammary carcinogenesis	(Xing et al., 1997)
Tumor cells inoculation (Mat B III-uPAR) in Fischer rats	ruPAR IgG	s.c. injection (50-100 µg/day for 7 days)	Inhibited mammary carcinogenesis	(Rabbani and Gladu, 2002)
Tumor cells inoculation (R3230) in Fischer 344 rats	Recombinant bone morphogenetic protein-2 (rBMP-2)	i.p. administration (10, 50 and 100 µg, single injection)	Dose-dependent calcifications were produced	(Liu et al., 2010)
Tumor cells (SST-2) inoculation in SHR female rats	Doxorubicin Mito-Tempol (Mito-T) Dexrazoxane	Doxorubicin (i.v., 10 mg/kg) Mito-T (i.p., 5 and 25 mg/kg) Dexrazoxane (i.p. 50 mg/kg) Mito-T + Dexrazoxane (i.p., 5 and 25 mg/kg + 50 mg/kg)	Doxorubicin inhibited mammary carcinogenesis Mito-T and Dexrazoxane inhibited mammary carcinogenesis and ameliorated doxorubicin-induced cardiomyopathy	(Dickey et al., 2013)
Tumor cells injection (s.c. or i.v., c-SST-2) in SHR rats	Malotilate	Oral administration (150 mg/kg b.w. for 7 days)	Lung metastases were inhibited	(Nagayasu et al., 1998)
Tumor cells (Walker 256) inoculation in Sprague-Dawley rats	Sunitinib malate Fingolimod	Sunitinib malate (oral gavage, 30 mg/kg for 5 or 7 days) Fingolimod (oral gavage, 5 mg/kg for 5 or 7 days) Sunitinib malate + Fingolimod (oral gavage, 30 mg/kg + 5 mg/kg for 5 or 7 days)	The drugs combination inhibited mammary carcinogenesis	(Mousseau et al., 2012)

Table 1. (Continued)

Model/gender and animal strain	Drugs and/or compounds evaluated	Dose/treatment	Therapeutic effects	Ref.
Tumor cells inoculation (BN472) in Brown-Norway	NVP-BEZ235	NVP-BEZ235 (oral administration, 5 mL/kg for 6 days)	Inhibited mammary carcinogenesis	(Schnell et al., 2008)
Genetically engineered model of mammary cancer				
MMTV-c-erbB-2 transgenic rat	-	Cystic expansions, sclerosing adenosis, and ductal hyperplasia were developed		(Davies et al., 1999)
MMTV-TGFα transgenic rat	-	Severe hyperplastic lumps, hyperplasia, papillary ductal adenoma, lactating adenoma, ductal carcinoma *in situ* and carcinoma were observed		(Davies et al., 1999)
Transgenic rats carrying human c-Ha-ras proto-oncogenes	*N*-methyl-*N*-nitrosourea (MNU)	i.v. (50 mg/kg at 50 days of age)	All MNU-exposed animals developed mammary tumors	(Asamoto et al., 2000)
c-Ha-ras transgenic rats non-transgenic rats	Purple corn color	Diet supplementation (5% for 8 weeks for transgenic animals and 22 weeks for non-transgenic ones)	Inhibited mammary carcinogenesis	(Fukamachi et al., 2008)
Transgenic rats carrying human c-Ha-ras proto-oncogenes + *N*-methyl-*N*-nitrosourea (MNU)	Isoflavones	Diet supplementation (0.25% for 20 and 56 days)	Inhibited mammary carcinogenesis	(Matsuoka et al., 2003)
Transgenic rats for *neu* proto-oncogene	Androgen 5α-dihydrotestosterone	For 6 months	Both females and males developed mammary carcinomas	(Watson et al., 2002)

Model/gender and animal strain	Drugs and/or compounds evaluated	Dose/treatment	Therapeutic effects	Ref.
WKAH and F344 strains carrying the human T-lymphotropic virus type I	-	Spontaneously developed mammary tumors at 5 months of age		(Yamada et al., 1995)
p53 knockout rats	Diethylnitrosamine	i.p. administration (20 mg/kg b.w., for 5 weeks)	Decreased survival time and latency period	(Yan et al., 2012)
Brca2 knockout rats	-	Underdeveloped mammary glands, cataract formation and short lifespan		(Cotroneo et al., 2007b)

b.w. body weight, i.p. intraperitoneal injection, i.v. intravenous injection, s.c. subcutaneous injection.

The simultaneous administration of antihistamines and grapefruit juice change their plasmatic concentration due to the blockage of cytochrome P450 (CYP) 3A4 (Bartra et al., 2006; Simons, 2004). The relation between the use of antihistaminic drugs and cancer risk development has intrigued the researchers. Indeed, it was observed in several studies that the aminoethyl ether group of antihistamines is structurally similar to N,N-diethyl-2-(4-(phenylmethyl)phenoxy) ethane-mine HCl (DPPE) that is a tamoxifen derivative known to inhibit the *in vitro* growth of MCF-7 breast cancer cells (Brandes et al., 1994; Brandes and Macdonald, 1985; Nadalin et al., 2003).

In this way, investigators have studied this association with distinct results (Cianchi, Vinci, and Masini, 2008). Nadalin and coworkers (Nadalin et al., 2003) enquired 3,133 women with breast cancer and 3,062 healthy women with age ranging from 25 to 74 years-old about the regular use of antihistamines, and they found no association between the antihistamines use and the risk of breast cancer development. Kelly and coworkers (Kelly et al., 1999) also studied the association between antihistamines and breast cancer risk in 5,814 women with invasive breast cancer and in 5,095 healthy women between 18 and 69 years of age, finding no association between antihistamines use and cancer development (Kelly et al., 1999).

Several researchers performed *in vitro* and *in vivo studies* with human cell lines of different types of cancer, namely leukemia, lymphoma, melanoma, breast, ovarian, vaginal, cervical, uterine, vulvar and colorectal cancer, demonstrating the positive involvement of histamine in cancer cell proliferation migration ad invasion (Medina et al., 2011).

In a study performed by our research team, where the role of mast cells was evaluated in the initiation and progression of mammary tumors chemically-induced by the carcinogen agent MNU in Sprague-Dawley female rats, through the inhibition of mast cell degranulation by the administration of ketotifen, we observed that animals from ketotifen-treated groups developed less number of mammary tumors (palpable masses) but higher number of mammary lesions when compared with non-treated animals. A lower proliferation (Ki-67 immunoexpression) and apoptotic index (caspase-3 and -9 immunoexpression) was observed in mammary tumors from ketotifen-exposed animals. The main positive effect of mast cell inhibition seemed to be the reduction of tumor proliferation when the mast cell degranulation was inhibited before tumor development (Faustino-Rocha et al., 2017).

TESTING THE ROLE OF LIFESTYLE ON MAMMARY CANCER DEVELOPMENT

The development of breast cancer is intimately associated with several risk factors, namely race, age, sex, reproductive factors, estrogens receptors, body weight, genetic mutations and lifestyle (Kamińska et al., 2015; Singletary, 2003).

The concept that breast cancer development may be preventable by lifestyle, namely physical activity, is supported by epidemiologic data worldwide (Coyle, 2008). Among the potential anticancer effects of physical activity are the decrease in endogenous sex hormone concentration and exposure (later age of menarche, decreased estrogen concentrations, decreased number of ovulatory cycles), favorable changes in body weight, insulin resistance, and chronic low-grade inflammation (Adams et al., 2006; Jones et al., 2005; McTiernan et al., 1998; Tehard et al., 2006; Westerlind et al., 2002). It was suggested that regular physical activity with a frequency of 3-5 times a week reinforces the immune system, reducing breast cancer risk by 20-40%. Moreover, physical activity improves general fitness and quality of life (Antoniou et al., 2003).

Our research team performed an experimental assay aiming to evaluate the effects of exercise training on mammary cancer development. For this, we used the rat model of mammary chemically-induced by the administration of the carcinogen agent MNU. The female Sprague-Dawley rats were exercised on a treadmill running (Treadmill Control® LE 8710, Panlab, Harvard Apparatus, Holliston, MA, US) at a speed of 20 m/min, 60 min/day, five times a weeks, for 35 weeks. The number of mammary tumors (23 versus 28, $p > 0.05$) and lesions (50 versus 71, $p = 0.056$) was lower in animals from exercised groups when compared with the sedentary ones. Exercised animals showed lower number of malignant lesions when compared with sedentary animals (21 versus 39, $p = 0.020$). Exercised animals presented lower serum levels of C-reactive protein, when compared with sedentary animals. Mammary tumors from exercised animals were more vascularized and exhibited higher expression of estrogen receptors α ($p < 0.05$) when compared with mammary tumors from sedentary animals. We concluded that lifelong endurance training has beneficial effects on mammary tumorigenesis in female rats (reduced the inflammation, the number of mammary tumors and lesions, and malignancy). Additionally, the mammary tumors from MNU exercised group exhibited higher immunoexpression of estrogen receptor α that is an indicator of well-differentiated tumors and better response to hormone therapy (Alvarado et al., 2016; Faustino-Rocha et al., 2016b, 2013c, 2015a, 2016f).

CONCLUSION

Experimental data related to the rat models of mammary cancer was reviewed in this chapter. Although several animal models are available for mammary cancer research, no one of them is perfect but they provide a starting point to study the Human mammary carcinogenesis and evaluate the potential role of new preventive and therapeutic strategies. The researchers should be able to select the model that best suits the aims of their studies after considering the advantages and disadvantages of each one. The model of chemically-induced mammary tumors in female rats is one of the most frequently used due to the its advantages when compared with other models, namely high incidence rate, short period of latency, high number of tumors, tumors similar to those found in humans in terms of histology, hormone dependency, expression of estrogen receptors and genetic alterations.

REFERENCES

Adams, SA, Matthews, CE, Hebert, JR, Moore, CG, Cunningham, JE, Shu, XO, Fulton, J, Gao, YT, Zheng, W. (2006). Association of physical activity with hormone receptor status: The Shanghai Breast Cancer Study. *Cancer Epidemiology Biomarkers & Prevention.* 15(6): 1170-8.

Ahelm, CN, Frincke, JM, White, SK, Reading, CL, TRauger, RJ, Lakshmanaswamy, R. (2011). 17α-ethynyl-5α-androstane-3α, 17β-diol treatment of MNU-induced mammary cancer in rats. *International Journal of Breast Cancer* 2011: 1-9.

Al-Dhaheri, WS, Hassouna, I, Al-Salam, S, Karam, SM. (2008). Characterization of breast cancer progression in the rat. *Recent Advances in Clinical Oncology* 1138: 121-31.

Al-Saeedi, FJ. (2014). Study of the cytotoxicity of asiaticoside on rats and tumour cells. *BMC cancer* 14: 220.

Alfaro, Y, Delgado, G, Cárabez, A, Anguiano, B, Aceves, C. (2013). Iodine and doxorubicin, a good combination for mammary cancer treatment: antineoplastic adjuvancy, chemoresistance inhibition, and cardioprotection. *Molecular Cancer* 12(45): 1-11.

Alshafie, GA, Walker, JR, Curley, RW, Clagett-Dame, M, Highland, MA, Nieves, NJ, Stonerock, LA, Abou-Issa, H. (2005). Inhibition of mammary tumor growth by a novel nontoxic retinoid: chemotherapeutic evaluation of a C-linked analog of 4-HPR-glucuronide. *Anticancer research* 25(3c): 2391-8.

Alvarado, A, Faustino-rocha, ANAI, Ferreira, R, Mendes, R, Duarte, JA, Pires, MJ, Colaço, B, Oliveira, PA. (2016). Prognostic factors in an exercised model of chemically-induced mammary cancer. *Anticancer Research* 36(5): 1-8.

Anbuselvam, C, Vijayavel, K, Balasubramanian, MP. (2007). Protective effect of Operculina turpethum against 7,12-dimethyl benz (a)anthracene induced oxidative stress with reference to breast cancer in experimental rats. *Chemico-Biological Interactions* 168(3): 229-36.

Andreea, G, Pegza, R, Lascu, L, Bondari, S, Stoica, Z, Bondari, A. (2011). The role of imaging techniques in diagnosis of breast cancer. *Current Health Sciences Journal* 37(2): 55-61.

Antoniou, A, Pharoah, PDP, Narod, S, Risch, HA, Eyfjord, JE, Hopper, JL, Loman, N, Olsson, H, Johannsson, O, Borg, A, Pasini, B, Radice, P, Manoukian, S, Eccles, DM, Tang, N, Olah, E, Anton-Culver, H, Warner, E, Lubinski, J. (2003). Average risks of breast and ovarian cancer associated with BRCA1 or BRCA2 mutations detected in case Series unselected for family history: a combined analysis of 22 studies. *American journal of human genetics* 72(5): 1117-30.

Anzano, MA, Peer, CW, Smith, JM, Mullen, LT, Shrader, MW, Logsdon, DL, Driver, CL, Brown, CC, Roberts, AB, Sporn, MB. (1996). Chemoprevention of mammary carcinogenesis in the rat: combined use of raloxifene and 9-cis-retinoic acid. *Journal of the National Cancer Institute.* 88(2): 123-25.

Arcos, J. (1995). *Chemical induction of cancer: modulation and combination effects.* Arcos J, Argus M, Woo Y-F (eds). Birkhäuser: Boston.

Arroyo-Acevedo, J, Chávez-Asmat, RJ, Anampa-Guzmán, A, Donaires, R, Ráez-Gonzáles, J. (2015). Protective Effect of Piper aduncum Capsule on DMBA-induced Breast Cancer in Rats. *Breast cancer: basic and clinical research* 9: 41-8.

Asamoto, M, Ochiya, T, Toriyama-Baba, H, Ota, T, Sekiya, T, Terada, M, Tsuda, H. (2000). Transgenic rats carrying human c-Ha-ras proto-oncogenes are highly susceptible to N-methyl-N-nitrosourea mammary carcinogenesis. *Carcinogenesis* 21(2): 243-9.

Avraham, R, Benish, M, Inbar, S, Bartal, I, Rosenne, E, Ben-Eliyahu, S. (2010). Synergism between immunostimulation and prevention of surgery-induced immune suppression: an approach to reduce post-operative tumor progression. *Brain, behavior, and immunity. NIH Public Access* 24(6): 952-8.

Avril, N, Menzel, M, Dose, J, Schelling, M, Weber, W, Jänicke, F, Nathrath, W, Schwaiger, M. (2001). Glucose metabolism of breast cancer assessed by 18F-FDG PET: histologic and immuno-histochemical tissue analysis. *Journal of nuclear medicine : official publication, Society of Nuclear Medicine* 42(1): 9-16.

Badawi, AF, Eldeen, MB, Liu, YY, Ross, EA, Badr, MZ. (2004). Inhibition of rat mammary gland carcinogenesis by simultaneous targeting of cyclooxygenase-2 and peroxisome proliferator-activated receptor gamma. *Cancer Research*. 64(3): 1181-9.

Badyal, DK, Desai, C. (2014). Animal use in pharmacology education and research: the changing scenario. *Indian journal of pharmacology*. Medknow Publications 46(3): 257-65.

Baek, H-M, Yu, HJ, Chen, J-H, Nalcioglu, O, Su, M-Y. (2008). Quantitative correlation between (1)H MRS and dynamic contrast-enhanced MRI of human breast cancer. Magnetic resonance imaging. *NIH Public Access* 26(4): 523-31.

Barros, AC, Muranaka, EN, Mori, LJ, Pelizon, HT, Iriya, K, Giocondo, G, Pinotti, JA. (2004). Induction of Experimental Mammary Carcinogenesis in Rats with 7,12-Dimethylbenz(a)Anthracene. *Revista do Hospital das Clínicas da Faculdade de Medicina da Universidade de São Paulo* 59(5): 257-61.

Bartella, L, Thakur, SB, Morris, EA, Dershaw, DD, Huang, W, Chough, E, Cruz, MC, Liberman, L. (2007). Enhancing nonmass lesions in the breast: evaluation with proton (1H) MR spectroscopy. *Radiology* 245(1): 80-7.

Bartra, J, Velero, AL, Del Cuvillo, A, Dávila, I, Jáuregui, I, Montoro, J, Mullol, J, Sastre, J. (2006). Interactions of the H1 antihistamines. *Journal of Allergy and Clinical Immunology* 16(1): 29-36.

Bartstra, RW, Bentvelzen, PAJ, Zoetelief, J, Mulder, AH, Broerse, JJ, van Bekkum, DW. (1998). Induction of mammary tumors in rats by single-dose gamma irradiation at different ages. *Radiation Research* 150(4): 442.

Batcioglu, K, Uyumlu, AB, Satilmis, B, Yildirim, B, Yucel, N, Demirtas, H, Onkal, R, Guzel, RM, Djamgoz, MBA. (2012). Oxidative stress in the in vivo DMBA rat model of breast cancer: suppression by a voltage-gated sodium channel inhibitor (RS100642). *Basic & clinical pharmacology & toxicology* 111(2): 137-41.

Benakanakere, I, Besch-Williford, C, Carroll, CE, Hyder, SM. (2010). Synthetic progestins differentially promote or prevent 7,12-dimethylbenz(a)anthracene-induced mammary tumors in sprague-dawley rats. *Cancer prevention research (Philadelphia, Pa.)* 3(9): 1157-67.

Bhowmik, A, Das, N, Pal, U, Mandal, M, Bhattacharya, S, Sarkar, M, Jaisankar, P, Maiti, NC, Ghosh, MK. (2013). 2,2'-diphenyl-3,3'-diindolylmethane: a potent compound induces apoptosis in breast cancer cells by inhibiting EGFR pathway. *PloS one* 8(3): 59798.

Białek, A, Tokarz, A, Zagrodzki, P. (2015). Conjugated linoleic acids (CLA) decrease the breast cancer risk in dmba-treated rats. *Acta poloniae pharmaceutica* 73(2): 315-27.

Bishayee, A, Mandal, A, Bhattacharyya, P, Bhatia, D. (2016). Pomegranate exerts chemoprevention of experimentally induced mammary tumorigenesis by suppression of cell proliferation and induction of apoptosis. *Nutrition and cancer* 68(1): 120-30.

Bobrowska-Korczak, B, Skrajnowska, D, Tokarz, A. (2012). The effect of dietary zinc--and polyphenols intake on DMBA-induced mammary tumorigenesis in rats. *Journal of biomedical science* 19: 43.

Boyland, E, Sydnor, K. (1962). The induction of mammary cancer in rats. *British Journal of Cancer* 16(4): 731-9.

Brandes, LJ, Warrington, RC, Arron, RJ, Bogdanovic, RP, Fang, W, Queen, GM, Stein, DA, Tong, JG, Zaborniak, CLF, Labella, FS.

(1994). Enhanced cancer growth in mice administered daily human-equivalent doses of some H1-antihistamines - predictive in vitro correlates. *Journal of the National Cancer Institute* 86(10): 770-5.

Brandes, LJ, Macdonald, LM. (1985). Evidence That the Antiestrogen Binding-Site Is A Histamine Or Histamine-Like Receptor. *Biochemical and Biophysical Research Communications* 126(2): 905-10.

Cardiff, RD. (2007). Epilog: comparative medicine, one medicine and genomic pathology. *Breast Diseases* 28: 107-10.

Carroll, CE, Liang, Y, Benakanakere, I, Besch-Williford, C, Hyder, SM. (2013). The anticancer agent YC-1 suppresses progestin-stimulated VEGF in breast cancer cells and arrests breast tumor development. *International journal of oncology.* 42(1): 179-87.

Carson, S, Lee, N, Thakurta, S. (2010). *Drug Class Review: Newer Antihistamines.* McDonagh M, Helfand M (eds). Oregon Health Science University: Oregon.

Carter, CL, Allen, C, Henson, DE. (1989). Relation of tumor size, lymph node status, and survival in 24,740 breast cancer cases. *Cancer* 63(1): 181-7.

Cerussi, AE, Jakubowski, D, Shah, N, Bevilacqua, F, Lanning, R, Berger, AJ, Hsiang, D, Butler, J, Holcombe, RF, Tromberg, BJ. (2002). Spectroscopy enhances the information content of optical mammography. *Journal of biomedical optics* 7(1): 60-71.

Chandra, M, Riley, MGI, Johnson, DE. (1992). Spontaneous Neoplasms in Aged Sprague-Dawley Rats. *Archives of Toxicology* 66(7): 496-502.

Chen, CP, Hanson, E, Watson, JW, Lee, JS. (2003). P-glycoprotein limits the brain penetration of nonsedating but not sedating H1-antagonists. *Drug Metabolism and Disposition* 31(3): 312-18.

Chun, YS, Bisht, S, Chenna, V, Pramanik, D, Yoshida, T, Hong, SM, de Wilde, RF, Zhang, Z, Huso, DL, Zhao, M, Rudek, MA, Stearns, V, Maitra, A, Sukumar, S. (2012). Intraductal administration of a polymeric nanoparticle formulation of curcumin (NanoCurc) significantly attenuates incidence of mammary tumors in a rodent chemical carcinogenesis model: Implications for breast cancer chemoprevention in at-risk populations. *Carcinogenesis* 33(11): 2242-49.

Cianchi, F, Vinci, MC, Masini, E. (2008). Histamine in cancer - the dual faces of the coin. *Cancer Biology & Therapy* 7(1): 36-7.

Clarke, R. (1996). Animal models of breast cancer: Their diversity and role in biomedical research. *Breast Cancer Research and Treatment* 39(1): 1-6.

Colditz, GA, Frazier, AL. (1995). Models of Breast-Cancer Show That Risk Is Set by Events of Early-Life - Prevention Efforts Must Shift Focus. *Cancer Epidemiology Biomarkers & Prevention* 4(5): 567-71.

Colston, KW, Pirianov, G, Bramm, E, Hamberg, J, Binderup, L. (2003). Effects of Seocalcitol (EB1089) on nitrosomethyl urea-induced rat mammary tumors. *Breast Cancer Research and Treatment* 80(3): 303-11.

Comission, E. (2013). *Seventh Report on the statistics on the number of animals used for experimental and other scientific purposes in the member states of the European Union.*

Committee for the Update of the Guide for the Care Animals, and U of L. (2011). *Guide for the Care and Use of Laboratory Animals.* The National Academies Press: Washighton D.C.

Conn, PM. (2013). Animal models for the study of human disease. *Animal Models for the Study of Human Disease*, Conn PM (ed). Elsevier Science & Technology.

Corsetti, V, Ferrari, A, Ghirardi, M, Bergonzini, R, Bellarosa, S, Angelini, O, Bani, C, Ciatto, S. (2006). Role of ultrasonography in detecting mammographically occult breast carcinoma in women with dense breasts. *La Radiologia medica* 111(3): 440-8.

Cortés-García, J, Aguilera-Méndez, A, Higareda-Mendoza, A, Beltrán-Peña, E, Pardo-Galván, M. (2009). Inducción de cáncer en rata wistar (Rattus norvegicus) mediante el uso de dimetil benzo(a)antraceno (DMBA) [Induction of Wistar rat cancer (Rattus norvegicus) by the use of dimethyl benzo (a) anthracene (DMBA)]. *EDEMM* 1(1): 11-17.

Cos, S, González, A, Güezmes, A, Mediavilla, MD, Martínez-Campa, C, Alonso-González, C, Sánchez-Barceló, EJ. (2006). Melatonin inhibits the growth of DMBA-induced mammary tumors by decreasing the local biosynthesis of estrogens through the modulation of aromatase activity. *International journal of cancer* 118(2): 274-8.

Cotroneo, MS, Haag, JD, Zan, Y, Lopez, CC, Thuwajit, P, Petukhova, G V, Camerini-Otero, RD, Gendron-Fitzpatrick, A, Griep, AE, Murphy, CJ, Dubielzig, RR, Gould, MN. (2007a). Characterizing a rat Brca2 knockout model. *Oncogene* 26(11): 1626-35.

Cotroneo, MS, Haag, JD, Zan, Y, Lopez, CC, Thuwajit, P, Petukhova, G V, Camerini-Otero, RD, Gendron-Fitzpatrick, A, Griep, AE, Murphy, CJ, Dubielzig, RR, Gould, MN. (2007b). Characterizing a rat Brca2 knockout model. *Oncogene*. Nature Publishing Group 26(11): 1626-35.

Coyle, YM. (2008). Physical activity as a negative modulator of estrogen-induced breast cancer. *Cancer Causes Control* 19: 1021-9.

Criado, PR, Criado, RFJ, Maruta, CW, Machado, CD. (2010). Histamine, histamine receptors and antihistamines: new concepts. *Anais Brasileiros de Dermatologia*. Univ Sao Paulo, Fac Med, Div Dermatol, Hosp Clin, Sao Paulo, Brazil ABC, Disciplina Dermatol, Fac Med, Sao Paulo, Brazil 85(2): 195-210.

Currier, N, Solomon, SE, Demicco, EG, Chang, DLF, Farago, M, Ying, HQ, Dominguez, I, Sonenshein, GE, Cardiff, RD, Xiao, ZXJ, Sherr, DH, Seldin, DC. (2005). Oncogenic signaling pathways activated in DMBA-induced mouse mammary tumors. *Toxicologic Pathology* 33(6): 726-37.

Das Gupta, S, So, JY, Wall, B, Wahler, J, Smolarek, AK, Sae-Tan, S, Soewono, KY, Yu, H, Lee, M-J, Thomas, PE, Yang, CS, Suh, N. (2015). Tocopherols inhibit oxidative and nitrosative stress in estrogen-induced early mammary hyperplasia in ACI rats. *Molecular Carcinogenesis* 54(9): 916-25.

Del Cuvillo, A, Mullol, J, Bartra, J, Dávilla, I, Jáuregui, I, Montoro, J, Sastre, J, Valero, AL. (2006). Comparative pharmacology of the H1 antihistamines. *Journal of Investigational Allergology and Clinical Immunology* 16(1): 3-12.

Davies, BR, Platt-Higgins, AM, Schmidt, G, Rudland, PS. (1999). Development of hyperplasias, preneoplasias, and mammary tumors in MMTV-c-erbB-2 and MMTV-TGFalpha transgenic rats. *The American journal of pathology*. American Society for Investigative Pathology 155(1): 303-14.

Deepalakshmi, K, Mirunalini, S. (2013). Modulatory effect of Ganoderma lucidum on expression of xenobiotic enzymes, oxidant-antioxidant and hormonal status in 7,12-dimethylbenz(a) anthracene-induced mammary carcinoma in rats. *Pharmacognosy magazine* 9(34): 167-75.

Deghan Manshadi, S, Ishiguro, L, Sohn, K-J, Medline, A, Renlund, R, Croxford, R, Kim, Y-I. (2014). Folic acid supplementation promotes mammary tumor progression in a rat model. *PloS one* 9(1): 84635.

Dickey, JS, Gonzalez, Y, Aryal, B, Mog, S, Nakamura, AJ, Redon, CE, Baxa, U, Rosen, E, Cheng, G, Zielonka, J, Parekh, P, Mason, KP, Joseph, J, Kalyanaraman, B, Bonner, W, Herman, E, Shacter, E, Rao, VA. (2013). Mito-tempol and dexrazoxane exhibit cardioprotective and chemotherapeutic effects through specific protein oxidation and autophagy in a syngeneic breast tumor preclinical model. *PloS one*. Public Library of Science 8(8): 70575.

Diwan, BA, Kasprzak, KS, Anderson, LM. (1997). Promotion of dimethylbenz[a]anthracene-initiated mammary carcinogenesis by iron in female Sprague-Dawley rats. *Carcinogenesis* 18(9): 1757-62.

Doctores, A, Martínez, J, Merchán, J, Sala, L, Renedo, G, Fernández-Pascual, J, Bullón, A. (1974). Carcinogenesis y Nitrosoamidas. *Patología* 7: 225-30.

Doyle, A, McGarry, MP, Lee, NA, Lee, JJ. (2012). The construction of transgenic and gene knockout/knockin mouse models of human disease. *Transgenic research* 21(2): 327–49. A Eickmeyer, SM, Gamble, GL, Shahpar, S, Do, KD. (2012). The role and efficacy of exercise in persons with cancer. *PM & R: The journal of injury, function, and rehabilitation* 4(11): 874-81.

Ericsson, AC, Crim, MJ, Franklin, CL. (2013). A brief history of animal modeling. *Missouri Medicine* 110(3): 201-5.

European Commission. (2010). *Of mice and men - are mice relevant models for human disease?* Outcomes of the European Commisiion worshop 'Are mice relevant models for human disease?' London, UK.

Fagundes, DJ, Taha, MO. (2004). Modelo animal de doença: critérios de escolha e espécies de animais de uso corrente [Animal disease model: selection criteria and species of animals in current use]. *Acta Cirúrgica Brasileira* 19(1): 59-65.

Fahey, JW, Zhang, Y, Talalay, P. (1997). Broccoli sprouts: an exceptionally rich source of inducers of enzymes that protect against chemical carcinogens. *Proceedings of the National Academy of Sciences of the United States of America* 94(19): 10367-72.

Faustino-Rocha, A, Gama, A, Neuparth, M, Oliveira, P, Ferreira, R, Ginja, M. (2017). Mast cells on mammary carcinogenesis: host or tumor supporters? *Anticancer Research* 37(3): 1013-21.

Faustino-Rocha, A, Oliveira, PA, Pinho-Oliveira, J, Teixeira-Guedes, C, Soares-Maia, R, da Costa, RG, Colaço, B, Pires, MJ, Colaço, J, Ferreira, R, Ginja, M. (2013a). Estimation of rat mammary tumor volume using caliper and ultrasonography measurements. *Lab Animal* 42(6): 217-24.

Faustino-Rocha, A, Gama, A, Oliveira, P, Alvarado, A, Ferreira, R, Ginja, M. (2016a). A spontaneous high-grade carcinoma in seven-week-old female rat. *Experimental and Toxicologic Pathology* 69(5):241-4.

Faustino-Rocha, A, Gama, A, Oliveira, PA, Alvarado, A, Neuparth, MJ, Ferreira, R, Ginja, M. (2016b). Effects of lifelong exercise training on mammary tumorigenesis induced by MNU in female Sprague-Dawley rats. *Clinical and Experimental Medicine* 17(2):151-60.

Faustino-Rocha, A, Gama, A, Oliveira, P, Ferreira, R, Ginja, M. (2016c). Spontaneous mammary tumor in a 7-week-old female rat. In *XXI Meeting of teh Portuguese Society of Animal Pathology*: 22-23.

Faustino-Rocha, AI, Gama, A, Oliveira, PA, Alvarado, A, Fidalgo-Gonçalves, L, Ferreira, R, Ginja, M. (2016). Ultrasonography as the gold standard for in vivo volumetric determination of chemically-induced mammary tumors. *In vivo* 30(4): 465-72.

Faustino-Rocha, AI, Silva, A, Gabriel, J, Reixeira-Guedes, C, Lopes, C, Gil da Costa, R, Gama, A, Ferreira, R, Oliveira, P, Ginja, M. (2013b). Ultrasonographic, thermographic and histologic evaluation of MNU-induced mammary tumors in female Sprague-Dawley rats. *Biomedicine & Pharmacotherapy* 67(8): 771-6.

Faustino-Rocha, AI, Silva, A, Gabriel, J, Teixeira-Guedes, CI, Lopes, C, da Costa, RG, Gama, A, Ferreira, R, Oliveira, PA, Ginja, M. (2013c). Ultrasonographic, thermographic and histologic evaluation of MNU-induced mammary tumors in female Sprague-Dawley rats. *Biomedicine & Pharmacotherapy*. 67(8): 771-6.

Faustino-Rocha, AI, Gama, A, Oliveira, PA, Alvarado, A, Vala, H, Ferreira, R, Ginja, M. (2015a). Expression of estrogen receptors-alpha and beta in chemicallyinduced mammary tumours. *Virchows Archiv*. CITAB, Dep Vet Sci, Vila Real, Portugal 467: 51.

Faustino-Rocha, AI, Silva, A, Gabriel, J, Gil da Costa, RM, Moutinho, M, Oliveira, PA, Gama, A, Ferreira, R, Ginja, M. (2016d). Long-term exercise training as a modulator of mammary cancer vascularization. *Biomedicine and Pharmacotherapy*. Elsevier Masson SAS 81: 273-80.

Faustino-Rocha, AI, Silva, A, Gabriel, J, Gil da Costa, R, Moutinho, M, Oliveira, P, Gama, A, Ferreira, R, Ginja, M. (2016e). Long-term exercise training as a modulator of mammary cancer vascularization. *Biomedicine & Pharmacotherapy* 81: 273-80.

Faustino-Rocha, AI, Silva, A, Gabriel, J, Gil da Costa, RM, Moutinho, M, Oliveira, PA, Gama, A, Ferreira, R, Ginja, M. (2016f). Long-term exercise training as a modulator of mammary cancer vascularization. *Biomedicine & pharmacotherapy* 81: 273-80.

Faustino-Rocha, AI. (2017). *Mammary carcinogenesis in female rats: contribution to monitoring and treatment*. Univeristy of Trás-os-Montes and Alto Douro.

Faustino-Rocha, AI, Ferreira, R, Oliveira, PA, Gama, A, Ginja, M. (2015b). N-methyl-N-nitrosourea as a mammary carcinogenic agent. *Tumour Biology* 36(12): 9095–117.

Faustino-Rocha, AI, Ferreira, R, Oliveira, PA, Gama, A, Ginja, M. (2015c). N-Methyl-N-nitrosourea as a mammary carcinogenic agent. *Tumor Biology*. 36(12): 9095-117.

Faustino-Rocha, AI, Oliveira, PA, Duarte, JA, Ferreira, R, Ginja, M. (2013d). Ultrasonographic Evaluation of Gastrocnemius Muscle in a Rat Model of N-Methyl-N-nitrosourea-induced Mammary Tumor. *In Vivo*. 27(6): 803-7.

Faustino-Rocha, AI, Pinto, C, Gama, A, Oliveira, PA. (2014). Effects of Ketotifen on Mammary Tumors Volume and Weight. *Anticancer Research* 34(10): 5902.

Fedrowitz, M, Kamino, K, Löscher, W. (2004). Significant differences in the effects of magnetic field exposure on 7,12-dimethylbenz(a)anthracene-induced mammary carcinogenesis in two substrains of Sprague-Dawley rats. *Cancer research* 64(1): 243-51.

Feig, SA, Shaber, GS, Schwartz, GF, Patchefsky, A, Libshitz, HI, Edeiken, J, Nerlinger, R, Curley, RF, Wallace, JD. (1977). Thermography, mammography, and clinical examination in breast cancer screening. Review of 16,000 studies. *Radiology* 122(1): 123-7.

Forabosco, F, Löhmus, M, Rydhmer, L, Sundström, LF. (2013). Genetically modified farm animals and fish in agriculture: A review. *Livestock Science* 153(1): 1-9.

Freudenstein, J, Dasenbrock, C, Nisslein, T. (2002). Lack of promotion of estrogen-dependent mammary gland tumors in vivo by an isopropanolic Cimicifuga racemosa extract. *Cancer research* 62(12): 3448-52.

Friedewald, SM, Rafferty, EA, Rose, SL, Durand, MA, Plecha, DM, Greenberg, JS, Hayes, MK, Copit, DS, Carlson, KL, Cink, TM, Barke, LD, Greer, LN, Miller, DP, Conant, EF. (2014). Breast cancer screening using tomosynthesis in combination with digital mammography. *JAMA* 311(24): 2499-507.

Fukamachi, K, Imada, T, Ohshima, Y, Xu, J, Tsuda, H. (2008). Purple corn color suppresses Ras protein level and inhibits 7,12-dimethylbenz[a]anthracene-induced mammary carcinogenesis in the rat. *Cancer Science* 99(9): 1841-6.

Gal, A, Baba, A, Miclaus, V, Bouari, C, Taulescu, M, Bolfa, P, Borza, G, Catoi, C. (2011). Comparative aspects regarding MNU-induced mammary carcinogenesis in immature Sprague-Dowley and Whistar rats. *Bulletin UASVM, Veterinary Medicine* 68(1): 159-63.

Ganta, C, Chiyo, D, Ayuzawa, R, Rachakatla, R, Pyle, M, Andrews, G, Weiss, M, Tamura, M, Troyer, D. (2009). Rat umbilical cord stem cells completely abolish rat mammary carcinomas with no evidence of metastasis or recurrence 100 days post-tumor cell inoculation. *Cancer Research* 69(5): 1815-20.

Garcia-Solis, P, Alfaro, Y, Anguiano, B, Delgado, G, Guzman, RC, Nandi, S, Diaz-Muñoz, M, Vazquez-Martinez, O, Aceves, C. (2005). Inhibition of N-methyl-N-nitrosourea-induced mammary carcinogenesis by molecular iodine(I-2) but not by iodide (I-) treatment evidence that I-2 prevents cancer promotion. *Molecular and Cellular Endocrinology* 236(1–2): 49-57.

Garcia-Solis, P, Yahia, EM, Aceves, C. (2008). Study of the effect of 'Ataulfo' mango (Mangifera indica L.) intake on mammary carcinogenesis and antioxidant capacity in plasma of N-methyl-N-nitrosourea (MNU)-treated rats. *Food Chemistry* 111(2): 309-15.

Gershon-Cohen, J, Haberman, JD. (1964). Thermography. *Radiology* 82: 280–5.

Giuseppetti, GM, Martegani, A, Di Cioccio, B, Baldassarre, S. (2005). Elastosonography in the diagnosis of the nodular breast lesions: preliminary report. *La Radiologia Medica* 110(1–2): 69-76.

Goldfarb, Y, Benish, M, Rosenne, E, Melamed, R, Levi, B, Glasner, A, Ben-Eliyahu, S. (2009). CpG-C oligodeoxynucleotides limit the deleterious effects of beta-adrenoceptor stimulation on NK cytotoxicity and metastatic dissemination. *Journal of immunotherapy* 32(3): 280-91.

Gordon, PB, Goldenberg, SL. (1995). Malignant breast masses detected only by ultrasound. A retrospective review. *Cancer* 76(4): 626–30.

Gottardis, MM, Jordan, C. (1987). Antitumor actions of keoxifene and tamoxifen in the N-nitrosomethylurea-induced rat mammary carcinoma model. *Cancer Research* 47: 4020-4.

Grubbs, CJ, Hill, DL, Bland, KI, Beenken, SW, Lin, TH, Eto, I, Atigadda, VR, Vines, KK, Brouillette, WJ, Muccio, DD. (2003). 9cUAB30, an RXR specific retinoid, and/or tarnoxifen in the prevention of methylnitrosourea-induced mammary cancers. *Cancer Letters* 201(1): 17-24.

Guo, Y, Mazar, AP, Lebrun, J-J, Rabbani, SA. (2002). An antiangiogenic urokinase-derived peptide combined with tamoxifen decreases tumor growth and metastasis in a syngeneic model of breast cancer. *Cancer Research* 62(16): 4678-84.

Haag, JD, Lindstrom, MJ, Gould, MN. (1992). Limonene-induced regression of mammary carcinomas. *Cancer Research* 52(14): 4021-6.

Hakkak, R, Shaaf, S, Jo, C-H, MacLeod, S, Korourian, S. (2010). Dehydroepiandrosterone intake protects against 7,12-dimethylbenz(a)anthracene-induced mammary tumor development in the obese Zucker rat model. *Oncology Reports* 24(2): 357–62.

Hakkak, R, Shaaf, S, Jo, CH, Macleod, S, Korourian, S. (2011). Effects of high-isoflavone soy diet vs. casein protein diet and obesity on DMBA-induced mammary tumor development. *Oncology letters.* Spandidos Publications 2(1): 29-36.

Hall, TJ, Zhu, Y, Spalding, CS. (2003). In vivo real-time freehand palpation imaging. *Ultrasound in Medicine & Biology* 29(3): 427-35.

Hamelin, BA, Bouayad, A, Drolet, B, Gravel, A, Turgeon, J. (1998). In vitro characterization of cytochrome P450 2D6 inhibition by classic histamine H-1 receptor antagonists. *Drug Metabolism and Disposition* 26(6): 536-9.

Harms, SE. (1998). Breast magnetic resonance imaging. *Seminars in ultrasound, CT, and MR* 19(1): 104-20.

Harvell, DME, Strecker, TE, Xie, B, Pennington, KL, McComb, RD, Shull, JD. (2002). Dietary energy restriction inhibits estrogen-induced mammary, but not pituitary, tumorigenesis in the ACI rat. *Carcinogenesis* 23(1): 161-9.

He, YU, Li, QQ, Guo, SC. (2016). Taurine attenuates Dimethylbenz [a]anthracene-induced breast tumorigenesis in rats: a plasma metabolomic study. *Anticancer research.* International Institute of Anticancer Research 36(2): 533-43.

Herbst, RS, Bajorin, DF, Bleiberg, H, Blum, D, Hao, D, Johnson, BE, Ozols, RF, Demetri, GD, Ganz, PA, Kris, MG, Levin, B, Markman, M, Raghavan, D, Reaman, GH, Sawaya, R, Schuchter, LM, Sweetenham, JW, Valat, LT, Vokes, EE, et al., (2006). Clinical cancer advances 2005: major research advances in cancer treatment, prevention, and screening - a report from the American Society of Clinical Oncology. *Journal of Clinical Oncology* 24(1): 190-205.

Hey, JA, Affrime, M, Cobert, B, Kreutner, W, Cuss, FM. (1999). Cardiovascular profile of loratadine. *Clinical and Experimental Allergy* 29: 197-9.

Hilakivi-Clarke, L, Onojafe, I, Raygada, M, Cho, E, Skaar, T, Russo, I, Clarke, R. (1999). Prepubertal exposure to zearalenone or genistein reduces mammary tumorigenesis. *British Journal of Cancer* 80(11): 1682-8.

Hinck, L, Silberstein, GB. (2005). Key stages in mammary gland development - The mammary end bud as a motile organ. *Breast Cancer Research* 7(6): 245-51.

Hoenerhoff, MJ, Shibata, MA, Bode, A, Green, JE. (2011). Pathologic progression of mammary carcinomas in a C3(1)/SV40 T/t-antigen transgenic rat model of human triple-negative and Her2-positive breast cancer. *Transgenic Research* 20(2): 247-59.

Holtzman, S, Stone, JP, Shellabarger, CJ. (1982). Radiation-Induced mammary carcinogenesis in virgin, pregnant, lactating, and post-lactating rats. *Cancer Research* 42(1): 50-3.

Howell, JS. (1963). Studies on chemically induced breast tumours in the rat. *Acta - Unio Internationalis Contra Cancrum* 19: 762-4.

Hvid, H, Thorup, I, Oleksiewicz, MB, Sjogren, I, Jensen, HE. (2011). An alternative method for preparation of tissue sections from the rat mammary gland. *Experimental and Toxicologic Pathology* 63(4): 317-24.

Hvid, H, Thorup, I, Sjogren, I, Oleksiewicz, MB, Jensen, HE. (2012). Mammary gland proliferation in female rats: effects of the estrous cycle, pseudo-pregnancy and age. *Experimental and Toxicologic Pathology* 64(4): 321-32.

Iannaccone, PM, Jacob, HJ. (2009a). Rats! *Disease Models & Mechanisms* 2(5–6): 206-10.

Iannaccone, PM, Jacob, HJ. (2009b). Rats! *Disease Models & Mechanisms* 2(5–6): 206-10.

Iatropoulos, MJ, Cerven, DR, de George, G, von Keutz, E, Williams, GM. (2008). Reduction by dietary matrix metalloproteinase inhibitor BAY 12-9566N of neoplastic development induced by diethylnitrosamine, N-nitrosodimethylamine, or 7,12-dimethylbenz (a)anthracene in rats. *Drug and Chemical Toxicology* 31(3): 305-16.

Ikezaki, S, Takagi, M, Tamura, K. (2011). Natural Occurrence of Neoplastic Lesions in Young Sprague-Dawley Rats. *Journal of Toxicologic Pathology* 24(1): 37-40.

Ip, C, Thompson, HJ, Zhu, Z, Ganther, HE. (2000). In vitro and in vivo studies of methylseleninic acid: evidence that a monomethylated selenium metabolite is critical for cancer chemoprevention. *Cancer research* 60(11): 2882-6.

Jain, AK, Thanki, K, Jain, S. (2013). Co-encapsulation of tamoxifen and quercetin in polymeric nanoparticles: implications on oral bioavailability, antitumor efficacy, and drug-induced toxicity. *Molecular Pharmaceutics* 10(9): 3459-74.

Jang, TJ, Jung, HG, Jung, KH, O, MK. (2002). Chemopreventive effect of celecoxib and expression of cyclooxygenase-1 and cyclooxy-genase-2 on chemically-induced rat mammary tumours. *International Journal of Experimental Pathology* 83(4): 173-82.

Jayakumar, JK, Nirmala, P, Praveen Kumar, BA, Kumar, AP. (2014). Evaluation of protective effect of myricetin, a bioflavonoid in dimethyl benzanthracene-induced breast cancer in female Wistar rats. *South Asian Journal of Cancer* 3(2): 107-11.

Jiang, S, Pogue, BW, Carpenter, CM, Poplack, SP, Wells, WA, Kogel, CA, Forero, JA, Muffly, LS, Schwartz, GN, Paulsen, KD, Kaufman, PA. (2009). Evaluation of breast tumor response to neoadjuvant chemotherapy with tomographic diffuse optical spectroscopy: case studies of tumor region-of-interest changes. *Radiology* 252(2): 551-60.

Joe, BN, Sickles, EA. (2014). The Evolution of Breast Imaging: Past to Present. *Radiology* 273(2): 23-44.

Jones, LW, Eves, ND, Courneya, KS, Chiu, BK, Baracos, VE, Hanson, J, Johnson, L, Mackey, JR. (2005). Effects of exercise training on antitumor efficacy of doxorubicin in MDA-MB-231 breast cancer xenografts. *Clinical Cancer Research* 11(18): 6695-98.

Kadir, EA, Sulaiman, SA, Yahya, NK, Othman, NH. (2013). Inhibitory effects of Tualang Honey on experimental breast cancer in rats: a preliminary study. *Asian Pacific Journal of Cancer Prevention* 14(4): 2249-54.

Kakehashi, A, Tago, Y, Yoshida, M, Sokuza, Y, Wei, M, Fukushima, S, Wanibuchi, H. (2012). Hormonally active doses of isoflavone aglycones promote mammary and endometrial carcinogenesis and alter the molecular tumor environment in Donryu rats. *Toxicological Sciences* 126(1): 39-51.

Kale, A, Gawande, S, Kotwal, S, Netke, S, Roomi, MW, Ivanov, V, Niedzwecki, A, Rath, M. (2010). A combination of green tea extract, specific nutrient mixture and quercetin: an effective intervention treatment for the regression of N-methyl-N-nitrosourea (MNU)-induced mammary tumors in Wistar rats. *Oncology Letters* 1(2): 313-17.

Kalender, WA, Beister, M, Boone, JM, Kolditz, D, Vollmar, S V., Weigel, MCC. (2012). High-resolution spiral CT of the breast at very low dose: concept and feasibility considerations. *European Radiology* 22(1): 1-8.

Kamińska, M, Ciszewski, T, Łopacka-Szatan, K, Miotła, P, Starosławska, E. (2015). Breast cancer risk factors. *Przegląd Menopauzalny* 14(3): 196-202.

Kaplan, SS. (2001). Clinical utility of bilateral whole-breast US in the evaluation of women with dense breast tissue. *Radiology* 221(3): 641-9.

Kararli, TT. (1995). Comparison of the gastrointestinal anatomy, physiology, and biochemistry of humans and commonly used laboratory animals. *Biopharmaceutics & Drug Disposition* 16(5): 351-80.

Karellas, A, Vedantham, S. (2008). Breast cancer imaging: a perspective for the next decade. *Medical Physics* 35(11): 4878-97.

Kaspareit, J, Rittinghausen, S. (1999). Spontaneous neoplastic lesions in Harlan Sprague-Dawley rats. *Experimental and Toxicologic Pathology*. 51(1): 105-7.

Kayashita, J, Shimaoka, I, Nakajoh, M, Kishida, N, Kato, N. (1999). Consumption of a buckwheat protein extract retards 7,12-dimethylbenz[alpha]anthracene-induced mammary carcinogenesis in rats. *Bioscience, Biotechnology, and Biochemistry* 63(10): 1837-9.

Kelly, JP, Rosenberg, L, Palmer, JR, Rao, RS, Strom, BL, Stolley, PD, Zauber, AG, Shapiro, S. (1999). Risk of breast cancer according to use of antidepressants, phenothiazines, and antihistamines. *American Journal of Epidemiology* 150(8): 861-8.

Khanna, C, Hunter, K. (2005). Modeling metastasis in vivo. *Carcinogenesis* 26(3): 513-23.

Kim, H, Hall, P, Smith, M, Kirk, M, Prasain, JK, Barnes, S, Grubbs, C. (2004a). Chemoprevention by grape seed extract and genistein in carcinogen-induced mammary cancer in rats is diet dependent. *The Journal of Nutrition* 134(12): 3445-52.

Kim, JB, O'Hare, MJ, Stein, R. (2004b). Models of breast cancer: is merging human and animal models the future? *Breast Cancer Research* 6(1): 22-30.

Kolb, TM, Lichy, J, Newhouse, JH. (1998). Occult cancer in women with dense breasts: detection with screening US-diagnostic yield and tumor characteristics. *Radiology* 207(1): 191-9.

Kolb, TM, Lichy, J, Newhouse, JH. (2002). Comparison of the performance of screening mammography, physical examination, and breast US and evaluation of factors that influence them: an analysis of 27,825 patient evaluations. *Radiology* 225(1): 165-75.

Kopans, DB. (2014). Digital breast tomosynthesis from concept to clinical care. *AJR. American Journal of Roentgenology* 202(2): 299-308.

Krishnamoorthy, D, Sankaran, M. (2016). Modulatory effect of Pleurotus ostreatus on oxidant/antioxidant status in 7, 12-dimethylbenz (a) anthracene induced mammary carcinoma in experimental rats - a dose-response study. *Journal of Cancer Research and Therapeutics* 12(1): 386-94.

Kristoffersen Wiberg, M, Aspelin, P, Perbeck, L, Boné, B. (2002). Value of MR imaging in clinical evaluation of breast lesions. *Acta Radiologica* 43(3): 275-81.

Krouskop, TA, Wheeler, TM, Kallel, F, Garra, BS, Hall, T. (1998). Elastic moduli of breast and prostate tissues under compression. *Ultrasonic Imaging* 20(4): 260-74.

Kubatka, P, Sadlonova, V, Kajo, K, Machalekova, K, Ostatnikova, D, Nosalova, G, Fetisovova, Z. (2008a). Neoplastic effects of exemestane in premenopausal breast cancer model. *Neoplasma* 55(6): 538-43.

Kubatka, P, Sadlonova, V, Kajo, K, Nosalova, G, Fetisovova, Z. (2008b). Preventive effects of letrozole in the model of premenopausal mammary carcinogenesis. *Neoplasma* 55(1): 42-6.

Kubota, K, Hisa, N, Ogawa, Y, Yoshida, S. (2002). Evaluation of tissue harmonic imaging for breast tumors and axillary lymph nodes. *Oncology reports* 9(6): 1335-8.

Kukreti, S, Cerussi, AE, Tanamai, W, Hsiang, D, Tromberg, BJ, Gratton, E. (2010). Characterization of metabolic differences between benign and malignant tumors: high-spectral-resolution diffuse optical spectroscopy. *Radiology* 254(1): 277-84.

Lamartiniere, CA, Jenkins, S, Betancourt, AM, Wang, J, Russo, J. (2011). Exposure to the Endocrine Disruptor Bisphenol A Alters Susceptibility for Mammary Cancer. *Hormone molecular biology and clinical investigation* 5(2): 45-52.

Lemon, HM, Kumar, PF, Peterson, C, Rodriguez-Sierra, JF, Abbo, KM. (1989). Inhibition of radiogenic mammary carcinoma in rats by estriol or tamoxifen. *Cancer* 63(9): 1685-92.

Li, JX, Cho, YY, Langfald, A, Carper, A, Lubet, RA, Grubbs, CJ, Ericson, ME, Bode, AM. (2011). Lapatinib, a preventive/therapeutic agent against mammary cancer, suppresses RTK-mediated signaling through multiple signaling pathways. *Cancer Prevention Research* 4(8): 1190-7.

Li, SA, Weroha, SJ, Tawfik, O, Li, JJ. (2002). Prevention of solely estrogen-induced mammary tumors in female aci rats by tamoxifen: evidence for estrogen receptor mediation. *The Journal of Endocrinology* 175(2): 297-305.

Lin, Y, Collier, AC, Liu, W, Berry, MJ, Panee, J. (2008). The inhibitory effect of bamboo extract on the development of 7,12-dimethylbenz[a]anthracene (DMBA)-induced breast cancer. *Phytotherapy Research* 22(11): 1440-5.

Lindfors, KK, Boone, JM, Newell, MS, D'Orsi, CJ. (2010). Dedicated breast computed tomography: the optimal cross-sectional imaging solution? *Radiologic Clinics of North America* 48(5): 1043-54.

Linjawi, SAA, Khalil, WKB, Hassanane, MM, Ahmed, ES. (2015). Evaluation of the protective effect of Nigella sativa extract and its primary active component thymoquinone against DMBA-induced breast cancer in female rats. *Archives of Medical Science* 11(1): 220-9.

Lippert, TH, Adlercreutz, H, Berger, MR, Seeger, H, Elger, W, Mueck, AO. (2003). Effect of 2-methoxyestradiol on the growth of methyl-nitroso-urea (MNU)-induced rat mammary carcinoma. *Journal of Steroid Biochemistry and Molecular Biology* 84(1): 51-6.

Liska, J, Macejova, D, Galbavy, S, Baranova, M, Zlatos, J, Stvrtina, S, Mostbock, S, Weiss, R, Scheiblhofer, S, Thalhamer, J, Brtko, J. (2003). Treatment of 1-methyl-1-nitrosourea-induced mammary tumours with immunostimulatory CpG motifs and 13-cis retinoic acid in female rats: histopathological study. *Experimental and Toxicologic Pathology* 55(2–3): 173-9.

Liska, J, Galbavy, S, Macejova, D, Zlatos, J, Brtko, J. (2000). Histopathology of mammary tumours in female rats treated with 1-Methyl-1-Nitrosourea. *Endocrine Regulations* 34: 91-6.

Liu, F, Misra, P, Lunsford, EP, Vannah, JT, Liu, Y, Lenkinski, RE, Frangioni, J V. (2010). A dose- and time-controllable syngeneic animal model of breast cancer microcalcification. *Breast Cancer Research and Treatment* 122(1): 87-94.

Liu, J-R, Sun, X-R, Dong, H-W, Sun, C-H, Sun, W-G, Chen, B-Q, Song, Y-Q, Yang, B-F. (2008). β-Ionone suppresses mammary carcinogenesis, proliferative activity and induces apoptosis in the mammary gland of the Sprague-Dawley rat. *International Journal of Cancer* 122(12): 2689-98.

Liu, JR, Dong, HW, Chen, BQ, Zhao, P, Liu, RH. (2009). Fresh apples suppress mammary carcinogenesis and proliferative activity and induce apoptosis in mammary tumors of the Sprague-Dawley rat. *Journal of Agricultural and Food Chemistry* 57(1): 297-304.

Lopes, AC, Cova, TFGG, Pais, AACC, Pereira, JLGFSC, Colaço, B, Cabrita, AMS. (2014). Improving discrimination in the grading of rat mammary tumors using two-dimensional mapping of histopathological observations. *Experimental and Toxicologic Pathology* 66(1): 73-80.

Lu, JX, Jiang, C, Mitrenga, T, Cutter, G, Thompson, HJ. (1998). Pathogenetic characterization of 1-methyl-1-nitrosourea-induced mammary carcinomas in the rat. *Carcinogenesis* 19(1): 223-7.

Lu, JX, Pei, HY, Kaeck, M, Thompson, HJ. (1997). Gene expression changes associated with chemically induced rat mammary carcinogenesis. *Molecular Carcinogenesis* 20(2): 204-15.

Lubet, RA, Steele, VE, Eto, I, Juliana, MM, Kelloff, GJ, Grubbs, CJ. (1997). Chemopreventive efficacy of anethole trithione, N-acetyl-L-cysteine, miconazole and phenethylisothiocyanate in the DMBA-induced rat mammary cancer model. *International Journal of Cancer* 72(1): 95-101.

Lubet, RA, Christov, K, Nunez, NP, Hursting, SD, Steele, VE, Juliana, MM, Eto, I, Grubbs, CJ. (2005). Efficacy of Targretin on methylnitrosourea-induced mammary cancers: prevention and therapy dose-response curves and effects on proliferation and apoptosis. *Carcinogenesis* 26(2): 441-8.

Lubet, RA, Heckman, BM, De Flora, SL, Steele, VE, Crowell, JA, Juliana, MM, Grubbs, CJ. (2011). Effects of 5,6-benzoflavone, indole-3-carbinol (I3C) and diindolylmethane (DIM) on chemically-induced mammary carcinogenesis: is DIM a substitute for I3C? *Oncology Reports* 26(3): 731-6.

Lucas, JN, Rudmann, DG, Credille, KM, Irizarry, AR, Peter, A, Snyder, PW. (2007). The rat mammary gland: Morphologic changes as an indicator of systemic hormonal perturbations induced by xenobiotics. *Toxicologic Pathology* 35(2): 199-207.

Luzhna, L, Kutanzi, K, Kovalchuk, O. (2015). Gene expression and epigenetic profiles of mammary gland tissue: Insight into the differential predisposition of four rat strains to mammary gland cancer. *Mutation Research-Genetic Toxicology and Environmental Mutagenesis* a 779: 39-56.

Ma, D-F, Katoh, R, Zhou, H, Wang, P-Y. (2007). Promoting effects of milk on the development of 7,12-dimethylbenz(a)anthracene (DMBA)-induced mammary tumors in rats. *Acta Histochemica et Cytochemica* 40(2): 61-7.

Ma, D, Zhang, Y, Yang, T, Xue, Y, Wang, P. (2014). Isoflavone intake inhibits the development of 7,12-dimethylbenz(a)anthracene (DMBA)-induced mammary tumors in normal and ovariectomized rats. *Journal of Clinical Biochemistry and Nutrition* 54(1): 31-8.

Maeda, K, Ohkura, S, Tsukamura, H. (2000). Physiology of reproduction. In *The laboratory rat*. Academic Press: UK.

Mafuvadze, B, Cook, M, Xu, Z, Besch-Williford, CL, Hyder, SM. (2013). Effects of dietary apigenin on tumor latency, incidence and multiplicity in a medroxyprogesterone acetate-accelerated 7,12-dimethylbenz(a)anthracene-induced breast cancer model. *Nutrition and Cancer* 65(8): 1184-91.

Mandal, A, Bhatia, D, Bishayee, A. (2013). Simultaneous disruption of estrogen receptor and Wnt/β-catenin signaling is involved in methyl amooranin-mediated chemoprevention of mammary gland carcinogenesis in rats. *Molecular and Cellular Biochemistry*. NIH Public Access 384(1–2): 239-50.

Mandal, A, Bishayee, A. (2015a). Mechanism of Breast Cancer Preventive Action of Pomegranate: Disruption of Estrogen Receptor and Wnt/β-catenin signaling pathways. *Molecules* 20(12): 22315-28.

Mandal, A, Bishayee, A. (2015b). Trianthema portulacastrum Linn. displays anti-inflammatory responses during chemically induced rat mammary tumorigenesis through simultaneous and differential regulation of NF-κB and Nrf2 signaling pathways. *International Journal of Molecular Sciences* 16(2): 2426-45.

Mandal, A, Bishayee, A. (2015c). Trianthema portulacastrum Linn. displays anti-inflammatory responses during chemically induced rat mammary tumorigenesis through simultaneous and differential regulation of NF-κB and Nrf2 signaling pathways. *International Journal of Molecular Sciences* 16(2): 2426-45.

Manivannan, S, Nelson, CM. (2012). Dynamics of branched tissue assembly. *Stem Cell Research and Therapy* 3(1757–6512): 42.

Manna, S, Das, S, Chatterjee, M, Janarthan, M, Chatterjee, M. (2011). Combined supplementation of vanadium and fish oil suppresses tumor growth, cell proliferation and induces apoptosis in DMBA-induced rat mammary carcinogenesis. *Journal of Cellular Biochemistry* 112(9): 2327-39.

Marchesi, V. (2013). Breast cancer: Stable breast cancer xenograft models. *Nature Reviews Clinical Oncology* 10(8): 426.

Martin, G, Melito, G, Rivera, E, Levin, E, Davio, C, Cricco, G, Andrade, N, Caro, R, Bergoc, R. (1996). Effect of tamoxifen on intraperitoneal N-nitroso-N-methylurea induced tumors. *Cancer Letters* 100(1–2): 227-34.

Matsuoka, Y, Fukamachi, K, Hamaguchi, T, Toriyama-Baba, H, Kawaguchi, H, Kusunoki, M, Yoshida, H, Tsuda, H. (2003). Rapid emergence of mammary preneoplastic and malignant lesions in human c-Ha-ras proto-oncogene transgenic rats: possible application for screening of chemopreventive agents. *Toxicologic Pathology* 31(6): 632-7.

McCarthy, BJ, Rankin, K, Il'yasova, D, Erdal, S, Vick, N, Ali-Osman, F, Bigner, DD, Davis, F. (2011). Assessment of type of allergy and antihistamine use in the development of glioma. *Cancer Epidemiology, Biomarkers and Prevention* 20(2): 370-8.

Mccormick, DL, Moon, RC. (1983). Inhibition of mammary carcinogenesis by Flurbiprofen, a non-steroidal anti-inflammatory agent. *British Journal of Cancer* 48(6): 859-61.

Mceuen, CS. (1938). Occurrence of cancer in rats treated with oestrone. *The American journal of cancer* 34: 184-95.

Mcgregor, DH, Land, CE, Choi, K, Tokuoka, S, Liu, PI, Wakabayashi, T, Beebe, GW. (1977). Breast-cancer incidence among atomic bomb survivors, Hiroshima and Nagasaki, 1950-69. *Journal of the National Cancer Institute*. 59(3): 799-811.

Mcmartin, DN, Sahota, PS, Gunson, DE, Hsu, HH, Spaet, RH. (1992). Neoplasms and related proliferative lesions in control Sprague-Dawley rats from carcinogenicity studies - historical data and diagnostic considerations. *Toxicologic Pathology* 20(2): 212-25.

McTiernan, A, Ulrich, C, Slate, S, Potter, J. (1998). Physical activity and cancer etiology: associations and mechanisms. *Cancer Causes & Control* 9(5): 487-509.

Medina, D. (2007). Chemical carcinogenesis of rat and mouse mammary glands. *Breast Diseases* 28: 63-8.

Medina, VA, Lamas, DJM, Brenzoni, PG, Massari, N, Carabajal, E, Rivera, ES. (2011). Histamine receptors as potential therapeutic targets for cancer drug development. In *drug development - a case study based insight into modern strategies*. InTech Europe: Rijeka, Croatia.

Mehta, R, Hawthorne, M, Uselding, L, Albinescu, D, Moriarty, R, Christov, K, Mehta, R. (2000). Prevention of N-methyl-N-nitro-sourea-induced mammary carcinogenesis in rats by 1 alpha-hydroxy-vitamin D-5. *Journal of the National Cancer Institute* 92(22): 1836-40.

Meisamy, S, Bolan, PJ, Baker, EH, Pollema, MG, Le, CT, Kelcz, F, Lechner, MC, Luikens, BA, Carlson, RA, Brandt, KR, Amrami, KK, Nelson, MT, Everson, LI, Emory, TH, Tuttle, TM, Yee, D, Garwood, M. (2005). Adding in vivo quantitative 1H MR spectroscopy to improve diagnostic accuracy of breast MR imaging: preliminary results of observer performance study at 4.0 T. *Radiology* 236(2): 465-75.

Mesia-Vela, S, Sanchez, RI, Reuhl, KR, Conney, AH, Kauffman, FC. (2006). Phenobarbital treatment inhibits the formation of estradiol-dependent mammary tumors in the August-Copenhagen Irish rat. *The Journal of pharmacology and experimental therapeutics* 317(2): 590-7.

Montano, MM, Chaplin, LJ, Deng, H, Mesia-Vela, S, Gaikwad, N, Zahid, M, Rogan, E. (2007). Protective roles of quinone reductase and tamoxifen against estrogen-induced mammary tumorigenesis. *Oncogene* 26(24): 3587–90.

Morris, EA. (2002). Breast cancer imaging with MRI. *Radiologic clinics of North America* 40(3): 443-66.

Moss, HA, Britton, PD, Flower, CD, Freeman, AH, Lomas, DJ, Warren, RM. (1999). How reliable is modern breast imaging in differentiating benign from malignant breast lesions in the symptomatic population? *Clinical Radiology* 54(10): 676-82.

Mousseau, Y, Mollard, S, Faucher-Durand, K, Richard, L, Nizou, A, Cook-Moreau, J, Baaj, Y, Qiu, H, Plainard, X, Fourcade, L, Funalot, B, Sturtz, FG. (2012). Fingolimod potentiates the effects of sunitinib malate in a rat breast cancer model. *Breast cancer research and treatment* 134(1): 31-40.

Mullins, LJ, Brooker, G, Mullins, JJ. (2002). Transgenesis in the rat. In *Transgenesis techniques: principles and protocols*. Humana Press: Cardiff.

Mundhe, NA, Kumar, P, Ahmed, S, Jamdade, V, Mundhe, S, Lahkar, M. (2015). Nordihydroguaiaretic acid ameliorates cisplatin induced nephrotoxicity and potentiates its anti-tumor activity in DMBA induced breast cancer in female Sprague–Dawley rats. *International Immunopharmacology* 28(1): 634-42.

Murata, N, Fujimori, S, Ichihara, Y, Sato, Y, Yamaji, T, Tsuboi, H, Uchida, M, Suzuki, H, Yamada, M, Oikawa, T, Nemoto, H, Nobuhiro, J, Choshi, T, Hibino, S. (2006). Synthesis and anti-tumor activity of a fluorinated analog of medroxyprogesterone acetate (MPA), 9α-fluoromedroxyprogesterone acetate (FMPA). *Chemical & Pharmaceutical Bulletin* 54(11): 1567-70.

Murray, TJ, Ucci, AA, Maffini, M V, Sonnenschein, C, Soto, AM. (2009). Histological analysis of low dose NMU effects in the rat mammary gland. *BMC Cancer* 9: 267.

Nadalin, V, Cotterchio, M, Kreiger, N. (2003). Antihistamine use and breast cancer risk. *International Journal of Cancer* 106(4): 566-8.

Nagayasu, H, Hamada, J, Kawano, T, Konaka, S, Nakata, D, Shibata, T, Arisue, M, Hosokawa, M, Takeichi, N, Moriuchi, T. (1998). Inhibitory effects of malotilate on invasion and metastasis of rat mammary carcinoma cells by modifying the functions of vascular endothelial cells. *British Journal of Cancer* 77(9): 1371-7.

Nakhate, KT, Kokare, DM, Singru, PS, Taksande, AG, Kotwal, SD, Subhedar, NK. (2010). Hypothalamic cocaine- and amphetamine-regulated transcript peptide is reduced and fails to modulate feeding behavior in rats with chemically-induced mammary carcinogenesis. *Pharmacology Biochemistry and Behavior* 97(2): 340-9.

Nandhakumar, R, Salini, K, Niranjali Devaraj, S. (2012). Morin augments anticarcinogenic and antiproliferative efficacy against 7,12-dimethylbenz(a)-anthracene induced experimental mammary carcinogenesis. *Molecular and Cellular Biochemistry* 364(1–2): 79-92.

Nandi, S, Guzman, RC, Yang, J. (1995). Hormones and mammary carcinogenesis in mice, rats, and humans - a unifying hypothesis. *Proceedings of the National Academy of Sciences of the United States of America* 92(9): 3650-7.

Negi, AK, Bhatnagar, A, Agnihotri, N. (2016). Fish oil augments cele-coxib mediated alteration in apoptotic pathway in the initiation phase of 7,12-dimethylbenz(α)anthracene-induced mammary carcinogenesis. *Biomedicine & Pharmacotherapy* 79: 9-16.

Nielsen, TS, Khan, G, Davis, J, Michels, KB, Hilakivi-Clarke, L. (2011). Prepubertal exposure to cow's milk reduces susceptibility to carcinogen-induced mammary tumorigenesis in rats. *International Journal of Cancer* 128(1): 12-20.

Nishikawa, RM, Mawdsley, GE, Fenster, A, Yaffe, MJ. (1987). Scanned-projection digital mammography. *Medical Physics* 14(5): 717-27.

Noble, RL, Cutts, JH. (1959). Mammary tumors of the rat - review. *Cancer Research* 19(11): 1125-39.

Noguchi, M, Minami, M, Yagasaki, R, Kinoshita, K, Earashi, M, Kitagawa, H, Taniya, T, Miyazaki, I. (1997). Chemoprevention of DMBA-induced mammary carcinogenesis in rats by low-dose EPA and DHA. *British Journal of Cancer* 75(3): 348-53.

Oishi, Y, Yoshizawa, K, Suzuki, J, Makino, N, Hase, K, Yamauchi, K, Tsubura, A. (1995). Spontaneously occurring mammary adenocarcinoma in a 10-wk-old female rat. *Toxicologic Pathology* 23(6): 696-700.

Okugawa, H, Yamamoto, D, Uemura, Y, Sakaida, N, Tanano, A, Tanaka, K, Kamiyama, Y. (2005). Effect of perductal paclitaxel exposure on the development of MNU-induced mammary carcinoma in female S-D rats. *Breast Cancer Research and Treatment* 91(1): 29-34.

Oliveira, PA, Colaço, A, Chaves, R, Guedes-Pinto, H, De-La-Cruz P, LF, Lopes, C. (2007). Chemical carcinogenesis. *Anais da Academia Brasileira de Ciências* 79(4): 593-616.

Oliveira, PA. (2016). Chemical carcinogens. In *Oxford Textbook of Oncology*. Oxford University Press: New York.

Oliveira, PA, Colaco, A, De la Cruz P, LF, Lopes, C. (2006a). Experimental bladder carcinogenesis-rodent models. *Experimental Oncology* 28(1): 2-11.

Oliveira, PA, Colaco, A, De la Cruz P, LF, Lopes, C. (2006b). Experimental bladder carcinogenesis-rodent models. *Experimental Oncology* 28(1): 2-11.

Orel, SG, Schnall, MD. (2001). MR imaging of the breast for the detection, diagnosis, and staging of breast cancer. *Radiology* 220(1): 13-30.

Ouhtit, A, Ismail, MF, Othman, A, Fernando, A, Abdraboh, ME, El-Kott, AF, Azab, YA, Abdeen, SH, Gaur, RL, Gupta, I, Shanmuganathan, S, Al-Farsi, YM, Al-Riyami, H, Raj, MHG. (2014). Chemoprevention of rat mammary carcinogenesis by spirulina. *The American Journal of Pathology* 184(1): 296-303.

Ozdemir, I, Selamoglu Talas, Z, Gul, M, Ates, B, Gok, Y, Esrefoglu, M, Yilmaz, I. (2006). Inhibition of DMBA induced rat mammary duct damage by novel synthetic organoselenium compounds. *Japanese Association for Laboratory Animal Science* 55(5): 449-55.

Partridge, SC, McDonald, ES. (2013). Diffusion weighted magnetic resonance imaging of the breast: protocol optimization, interpretation, and clinical applications. *Magnetic Resonance Imaging Clinics of North America* 21(3): 601-24.

Periyasamy, K, Baskaran, K, Ilakkia, A, Vanitha, K, Selvaraj, S, Sakthisekaran, D. (2015). Antitumor efficacy of tangeretin by targeting the oxidative stress mediated on 7,12-dimethylbenz(a) anthracene-induced proliferative breast cancer in Sprague-Dawley rats. *Cancer Chemotherapy and Pharmacology* 75(2): 263-72.

Perletti, G, Concari, P, Giardini, R, Marras, E, Piccinini, F, Folkman, J, Chen, L. (2000). Antitumor activity of endostatin against carcinogen-induced rat primary mammary tumors. *Cancer Research* 60(7): 1793-6.

Pharmacists, AS of H-S. (2014). *Antihistamine Drugs.*

Pogue, BW, Poplack, SP, McBride, TO, Wells, WA, Osterman, KS, Osterberg, UL, Paulsen, KD. (2001). Quantitative hemoglobin tomography with diffuse near-infrared spectroscopy: pilot results in the breast. *Radiology* 218(1): 26-6.

Pugalendhi, P, Manoharan, S, Suresh, K, Baskaran, N. (2011). Genistein and daidzein, in combination, protect cellular integrity during 7,12-dimethylbenz[a]anthracene (DMBA) induced mammary carcinogenesis in Sprague-Dawley rats. *African Journal Of Traditional, Complementary, and Alternative Medicines* 8(2): 91-7.

Purushothaman, A, Nandhakumar, E, Sachdanandam, P. (2013). Phytochemical analysis and anticancer capacity of Shemamruthaa, a herbal formulation against DMBA- induced mammary carcinoma in rats. *Asian Pacific Journal of Tropical Medicin* 6(12): 925-33.

Qin, L-Q, Xu, J-Y, Wang, P-Y, Ganmaa, D, Li, J, Wang, J, Kaneko, T, Hoshi, K, Shirai, T, Sato, A. (2004). Low-fat milk promotes the development of 7,12-dimethylbenz(A)anthracene (DMBA)-induced mammary tumors in rats. *International Journal of Cancer* 110(4): 491-6.

Qin, L-Q, Xu, J-Y, Tezuka, H, Wang, P-Y, Hoshi, K. (2007). Commercial soy milk enhances the development of 7,12-dimethyl-benz(a)anthracene-induced mammary tumors in rats. *In vivo* 21(4): 667-71.

Quagliata, L, Klusmeier, S, Cremers, N, Pytowski, B, Harvey, A, Pettis, RJ, Thiele, W, Sleeman, JP. (2014). Inhibition of VEGFR-3 activation in tumor-draining lymph nodes suppresses the outgrowth of lymph node metastases in the MT-450 syngeneic rat breast cancer model. *Clinical & Experimental Metastasis* 31(3): 351-65.

Rabbani, SA, Gladu, J. (2002). Urokinase receptor antibody can reduce tumor volume and detect the presence of occult tumor metastases in vivo. *Cancer Research* 62(8): 2390-7.

Radisky, DC, Hirai, Y, Bissell, MJ. (2003). Delivering the message: epimorphin and mammary epithelial morphogenesis. *Trends in Cell Biology* 13: 426-34.

Rajakumar, T, Pugalendhi, P, Thilagavathi, S. (2015). Dose response chemopreventive potential of allyl isothiocyanate against 7,12-dimethylbenz(a)anthracene induced mammary carcinogenesis in female Sprague-Dawley rats. *Chemico-Biological Interactions* 231: 35-43.

Ravoori, S, Vadhanam, M V, Sahoo, S, Srinivasan, C, Gupta, RC. (2007). Mammary tumor induction in ACI rats exposed to low levels of 17 beta-estradiol. *International Journal of Oncology* 31(1): 113-20.

Rennó, AL, Alves-Júnior, MJ, Rocha, RM, De Souza, PC, de Souza, VB, Jampietro, J, Vassallo, J, Hyslop, S, Anhê, GF, de Moraes Schenka, NG, Soares, FA, Schenka, AA. (2015). Decreased expression of stem cell markers by simvastatin in 7,12-dimethylbenz(a)anthracene (DMBA)-induced breast cancer. *Toxicologic Pathology* 43(3): 400-10.

Rimmer, SJ, Church, MK. (1990). The Pharmacology and Mechanisms of Action of Histamine-H1-Antagonists. *Clinical and Experimental Allergy* 20: 3-17.

Roomi, MW, Roomi, NW, Ivanov, V, Kalinovsky, T, Niedzwiecki, A, Rath, M. (2005). Modulation of N-methyl-N-nitrosourea induced mammary tumors in Sprague-Dawley rats by combination of lysine, proline, arginine, ascorbic acid and green tea extract. *Breast Cancer Research* 7(3): 291-95.

Roy, P, George, J, Srivastava, S, Tyagi, S, Shukla, Y. (2011). Inhibitory effects of tea polyphenols by targeting cyclooxygenase-2 through regulation of nuclear factor kappa B, Akt and p53 in rat mammary tumors. *Investigational New Drugs* 29(2): 225-31.

Rudmann, D, Cardiff, R, Chouinard, L, Goodman, D, Kuttler, K, Marxfeld, H, Molinolo, A, Treumann, S, Yoshizawa, K. (2012). Proliferative and nonproliferative lesions of the rat and mouse mammary, zymbal's, preputial, and clitoral glands. *Toxicologic Pathology* 40(6): 7-39.

Ruhlen, RL, Willbrand, DM, Besch-Williford, CL, Ma, L, Shull, JD, Sauter, ER. (2009). Tamoxifen induces regression of estradiol-induced mammary cancer in the ACI.COP-Ept2 rat model. *Breast Cancer Research and Treatment* 117(3): 517-24.

Russo, IH, Russo, J. (1998). Role of hormones in mammary cancer initiation and progression. *Journal of Mammary Gland Biology and Neoplasia* 3(1): 49-61.

Russo, J, Russo, IH. (1996). Experimentally induced mammary tumors in rats. *Breast Cancer Research and Treatment* 39(1): 7-20.

Russo, J, Russo, IH. (2000). Atlas and histologic classification of tumors of the rat mammary gland. *Journal of Mammary Gland Biology and Neoplasia* 5(2): 187-200.

Russo, J, Tay, L, Russo, I. (1982). Differentiation of the mammary gland and susceptibility to carcinogenesis. *Breast Cancer Research and Treatment* 2: 5-73.

Saarinen, NM, Huovinen, R, Wärri, A, Mäkelä, SI, Valentín-Blasini, L, Sjöholm, R, Ammälä, J, Lehtilä, R, Eckerman, C, Collan, YU, Santti, RS. (2002). Enterolactone inhibits the growth of 7,12-dimethylbenz (a)anthracene-induced mammary carcinomas in the rat. *Molecular Cancer Therapeutics* 1(10): 86-76.

Sahin, K, Tuzcu, M, Sahin, N, Akdemir, F, Ozercan, I, Bayraktar, S, Kucuk, O. (2011). Inhibitory effects of combination of lycopene and genistein on 7,12- dimethyl benz(a)anthracene-induced breast cancer in rats. *Nutrition and Cancer* 63(8): 1279-86.

Samy, RP, Rajendran, P, Li, F, Anandi, NM, Stiles, BG, Ignacimuthu, S, Sethi, G, Chow, VTK. (2012). Identification of a novel Calotropis procera protein that can suppress tumor growth in breast cancer through the suppression of NF-κB pathway. *PloS One* 7(12): 48514.

Sano, D, Myers, JN. (2009). Xenograft models of head and neck cancers. *Head & Neck Oncology*. 1: 32.

Saraydin, SU, Tuncer, E, Tepe, B, Karadayi, S, Özer, H, Şen, M, Karadayi, K, Inan, D, Elagöz, Ş, Polat, Z, Duman, M, Turan, M. (2012). Antitumoral effects of Melissa officinalis on breast cancer in vitro and in vivo. *Asian Pacific Journal of Cancer Prevention* 13(6): 2765-70.

Sato, M, Pei, RJ, Yuri, T, Danbara, N, Nakane, Y, Tsubura, A. (2003). Prepubertal resveratrol exposure accelerates N-methyl-N-nitrosourea-induced mammary carcinoma in female Sprague-Dawley rats. *Cancer Letters* 202(2): 137-45.

Schaffer, EM, Liu, JZ, Green, J, Dangler, CA, Milner, JA. (1996). Garlic and associated allyl sulfur components inhibit N-methyl-N-nitrosourea induced rat mammary carcinogenesis. *Cancer Letters* 102(1–2): 199-204.

Schnell, CR, Stauffer, F, Allegrini, PR, O'Reilly, T, McSheehy, PMJ, Dartois, C, Stumm, M, Cozens, R, Littlewood-Evans, A, García-Echeverría, C, Maira, S-M. (2008). Effects of the dual phosphatidylinositol 3-kinase/mammalian target of rapamycin inhibitor NVP-BEZ235 on the tumor vasculature: implications for clinical imaging. *Cancer Research* 68(16): 6598-607.

Schuurman, H-J, Hougen, HP, Van Loveren, H. (1992). The rnu (Rowett Nude) and rnuN (nznu, New Zealand Nude) rat: an update. *ILAR Journal* 34: 1-2.

Sengupta, P. (2013). The Laboratory Rat: Relating Its Age With Human's. *International Journal of Preventive Medicine* 4(6): 624-30.

Shaban, N, Abdel-Rahman, S, Haggag, A, Awad, D, Bassiouny, A, Talaat, I. (2016). Combination between taxol-encapsulated liposomes and eruca sativa seed extract suppresses mammary tumors in female rats induced by 7,12 Dimethylbenz(α)anthracene. *Asian Pacific Journal of Cancer Prevention* 17(1): 117-23.

Shellabarger, CJ, Stone, JP, Holtzman, S. (1983). Effect of Interval Between Neutron Radiation and Diethylstilbestrol on Mammary Carcinogenesis in Female Acl Rats. *Environmental Health Perspectives* 50: 227-32.

Shivapurkar, N, Tang, ZC, Frost, A, Alabaster, O. (1996). A rapid dual organ rat carcinogenesis bioassay for evaluating the chemoprevention of breast and colon cancer. *Cancer Letters* 100(1–2): 169-79.

Shull, JD, Spady, TJ, Snyder, MC, Johansson, SL, Pennington, KL. (1997). Ovary-intact, but not ovariectomized female ACI rats treated with 17 beta estradiol rapidly develop mammary carcinoma. *Carcinogenesis* 18(8): 1595-601.

Siddiqui, IA, Sanna, V, Ahmad, N, Sechi, M, Mukhtar, H. (2015). Resveratrol nanoformulation for cancer prevention and therapy. *Annals of the New York Academy of Sciences* 1348: 20-31.

Silva, JC, Ferreira-Strixino, J, Fontana, LC, Paula, LM, Raniero, L, Martin, AA, Canevari, RA. (2014). Apoptosis-associated genes related to photodynamic therapy in breast carcinomas. *Lasers in Medical Science* 29(4): 1429-36.

Simons, FER. (2004). Drug therapy - Advances in H-1-antihistamines. *New England Journal of Medicine* 351(21): 2203-17.

Simons, FER, Silver, NA, Gu, XC, Simons, KJ. (2002). Clinical pharmacology of H-1-antihistamines in the skin. *Journal of Allergy and Clinical Immunology* 110(5): 777-83.

Singh, B, Shoulson, R, Chatterjee, A, Ronghe, A, Bhat, NK, Dim, DC, Bhat, HK. (2014). Resveratrol inhibits estrogen-induced breast carcinogenesis through induction of NRF2-mediated protective pathways. *Carcinogenesis* 35(8): 1872-80.

Singh, B, Bhat, NK, Bhat, HK. (2012). Induction of NAD(P)H-quinone oxidoreductase 1 by antioxidants in female ACI rats is associated with decrease in oxidative DNA damage and inhibition of estrogen-induced breast cancer. *Carcinogenesis* 33(1): 156-63.

Singletary, K, MacDonald, C, Iovinelli, M, Fisher, C, Wallig, M. (1998). Effect of the beta-diketones diferuloylmethane (curcumin) and dibenzoylmethane on rat mammary DNA adducts and tumors induced by 7,12-dimethylbenz[a]anthracene. *Carcinogenesis* 19(6): 1039-43.

Singletary, K, MacDonald, C, Wallig, M. (1997). The plasticizer benzyl butyl phthalate (BBP) inhibits 7,12-dimethylbenz[a]anthracene (DMBA)-induced rat mammary DNA adduct formation and tumorigenesis. *Carcinogenesis* 18(8): 1669-73.

Singletary, SE. (2003). Rating the risk factors for breast cancer. *Annals of surgery*. Lippincott, Williams, and Wilkins 237(4): 474-82.

Skrajnowska, D, Bobrowska-Korczak, B, Tokarz, A, Bialek, S, Jezierska, E, Makowska, J. (2013). Copper and resveratrol attenuates serum catalase, glutathione peroxidase, and element values in rats with DMBA-induced mammary carcinogenesis. *Biological Trace Element Research*r 156: 271-8.

Smits, BMG, Cotroneo, MS, Haag, JD, Gould, MN. (2007). Genetically engineered rat models for breast cancer. *Breast Disease* 28: 53-61.

Son, WC, Bell, D, Taylor, I, Mowat, V. (2010). Profile of Early Occurring Spontaneous Tumors in Han Wistar Rats. *Toxicologic Pathology* 38(2): 292-6.

Song, LL, Kosmeder, JW, Lee, SK, Gerhäuser, C, Lantvit, D, Moon, RC, Moriarty, RM, Pezzuto, JM. (1999). Cancer chemopreventive activity mediated by 4'-bromoflavone, a potent inducer of phase II detoxification enzymes. *Cancer Research* 59(3): 578-85.

Soriano, O, Delgado, G, Anguiano, B, Petrosyan, P, Molina-Servín, ED, Gonsebatt, ME, Aceves, C. (2011). Antineoplastic effect of iodine and iodide in dimethylbenz[a]anthracene-induced mammary tumors: association between lactoperoxidase and estrogen-adduct production. *Endocrine-Related Cancer* 18(4): 529-39.

Srivastava, P, Russo, J, Russo, IH. (1997). Chorionic gonadotropin inhibits rat mammary carcinogenesis through activation of programmed cell death. *Carcinogenesis* 18(9): 1799-808.

Stearns, V, Mori, T, Jacobs, LK, Khouri, NF, Gabrielson, E, Yoshida, T, Kominsky, SL, Huso, DL, Jeter, S, Powers, P, Tarpinian, K, Brown, RJ, Lange, JR, Rudek, MA, Zhang, Z, Tsangaris, TN, Sukumar, S. (2011). Preclinical and clinical evaluation of intraductally administered agents in early breast cancer. *Science Translational Medicine* 3(106): 106-8.

Steele, VE, Lubet, RA. (2010). The use of animal models for cancer chemoprevention drug development. *Seminars in Oncology* 37(4): 327-38.

Sternlicht, MD. (2006). Key stages in mammary gland development: the cues that regulate ductal branching morphogenesis. *Breast Cancer Research* 8(1): 201.

Sudo, K, Monsma, FJ, Katzenellenbogen, BS. (1983). Antiestrogen-binding sites distinct from the estrogen-receptor - subcellular localization, ligand specificity, and distribution in tissues of the Rat. *Endocrinology* 112(2): 425-34.

Sukumar, S, Notario, V, Martinzanca, D, Barbacid, M. (1983). Induction of mammary carcinomas in rats by nitroso-methylurea involves malignant activation of H-Ras-1 locus by single point mutations. *Nature* 306(5944): 658-61.

Szaefer, H, Krajka-Kuźniak, V, Ignatowicz, E, Adamska, T, Markowski, J, Baer-Dubowska, W. (2014). The effect of cloudy apple juice on hepatic and mammary gland phase I and II enzymes induced by DMBA in female Sprague-Dawley rats. *Drug and Chemical Toxicology* 37(4): 472-9.

Tabaczar, S, Domeradzka, K, Czepas, J, Piasecka-Zelga, J, Stetkiewicz, J, Gwoździński, K, Koceva-Chyła, A. (2015). Anti-tumor potential of nitroxyl derivative Pirolin in the DMBA-induced rat mammary carcinoma model: a comparison with quercetin. *Pharmacological Reports* 67(3): 527-34.

Tardivon, A, El Khoury, C, Thibault, F, Wyler, A, Barreau, B, Neuenschwander, S. (2007). [Elastography of the breast: a prospective study of 122 lesions]. *Journal de Radiologie* 88: 657-62.

Tehard, B, Friedenreich, CM, Oppert, JM, Clavel-Chapelon, F. (2006). Effect of physical activity on women at increased risk of breast cancer: results from the E3N cohort study. *Cancer Epidemiology Biomarkers & Prevention* 15(1): 57-64.

Tepsuwan, A, Kupradinun, P, Kusamran, WR. (2002). Chemopreventive potential of neem flowers on carcinogen-induced rat mammary and liver carcinogenesis. *Asian Pacific Journal of Cancer Prevention* 3(3): 231-38.

Thiele, W, Rothley, M, Teller, N, Jung, N, Bulat, B, Plaumann, D, Vanderheiden, S, Schmaus, A, Cremers, N, Göppert, B, Dimmler, A, Eschbach, V, Quagliata, L, Thaler, S, Marko, D, Bräse, S, Sleeman, JP. (2013). Delphinidin is a novel inhibitor of lymphangiogenesis but promotes mammary tumor growth and metastasis formation in syngeneic experimental rats. *Carcinogenesis* 34(12): 2804-13.

Thompson, HJ, Singh, M. (2000). Rat models of premalignant breast disease. *Journal of Mammary Gland Biology and Neoplasia* 5(4): 409-20.

Thompson, MD, Thompson, HJ, McGinley, JN, Neil, ES, Rush, DK, Holm, DG, Stushnoff, C. (2009). Functional food characteristics of potato cultivars (Solanum tuberosum L.): phytochemical composition and inhibition of 1-methyl-1-nitrosourea induced breast cancer in rats. *Journal of Food Composition and Analysis* 22(6): 571-6.

Thordarson, G, Lee, A V, McCarty, M, Van Horn, K, Chu, O, Chou, YC, Yang, J, Guzman, RC, Nandi, S, Talamantes, F. (2001). Growth and characterization of N-methyl-N-nitrosourea-induced mammary tumors in intact and ovariectomized rats. *Carcinogenesis* 22(12): 2039-48.

Turan, VK, Sanchez, RI, Li, JJ, Li, SA, Reuhl, KR, Thomas, PE, Conney, AH, Gallo, MA, Kauffman, FC, Mesia-Vela, S. (2004). The effects of steroidal estrogens in ACI rat mammary carcinogenesis: 17 beta-estradiol, 2-hydroxyestradiol, 4-hydroxyestradiol, 16 alpha-hydroxyestradiol, and 4-hydroxyestrone. *Journal of Endocrinology* 183(1): 91-9.

Ueda, M, Imai, T, Takizawa, T, Onodera, H, Mitsumori, K, Matsui, T, Hirose, M. (2005). Possible enhancing effects of atrazine on growth of 7,12-dimethylbenz(a) anthracene-induced mammary tumors in ovariectomized Sprague-Dawley rats. *Cancer Science* 96(1): 19-25.

Van der Gulden, WJI, Beynen, AC, Hau, J. (1999). Modelos Animales [Animal Models]. In *Principios de la Ciencia del Animals de Laboratorio* [Principles of Laboratory Animal Science], Van

Zutphen LFM, Baumans V, Beynen AC (eds). Elsevier: Granada: 211-9.

Van Schoor, J. (2012). Antihistamines: a brief review. *Professional Nursing Today* 16(5): 16-21.

Vandamme, TF. (2014). Use of rodents as models of human diseases. *Journal of Pharmacy & Bioallied Sciences* 6(1): 2-9.

Vanitha, MK, Priya, KD, Baskaran, K, Periyasamy, K, Saravanan, D, Venkateswari, R, Mani, BR, Ilakkia, A, Selvaraj, S, Menaka, R, Geetha, M, Rashanthy, N, Anandakumar, P, Sakthisekaran, D. (2015). Taurine regulates mitochondrial function during 7,12-Dimethyl Benz[a]anthracene induced experimental mammary carcinogenesis. *Journal of Pharmacopuncture* 18(3): 68-74.

Vargo-Gogola, T, Rosen, JM. (2007). Modelling breast cancer: one size does not fit all. *Nature Reviews Cancer* 7(9): 659-72.

Veena, K, Shanthi, P, Sachdanandam, P. (2006). Anticancer effect of Kalpaamruthaa on mammary carcinoma in rats with reference to glycoprotein components, lysosomal and marker enzymes. *Biological & Pharmaceutical Bulletin* 29(3): 565-9.

Vinodhini, J, Sudha, S. (2013). Effect of bis-carboxy ethyl germanium sesquoxide on N-nitroso-N-methylurea-induced rat mammary carcinoma. *Asian Journal of Pharmaceutical and Clinical Research* 6(2): 242-44.

Wagner, KU. (2004). Models of breast cancer: quo vadis, animal modeling? *Breast Cancer Research* 6(1): 31-8.

Watson, PA, Kim, K, Chen, K-S, Gould, MN. (2002). Androgen-dependent mammary carcinogenesis in rats transgenic for the Neu proto-oncogene. *Cancer Cell* 2(1): 67-79.

Westerlind, KC, McCarty, HL, Gibson, KJ, Strange, R. (2002). Effect of exercise on the rat mammary gland: implications for carcinogenesis. *Acta Physiologica Scandinavica* 175(2): 147-56.

Wiehle, RD, Christov, K, Mehta, R. (2007). Anti-progestins suppress the growth of established tumors induced by 7,12-dimethylbenz (a)anthracene: comparison between RU486 and a new 21-substituted-19-nor-progestin. *Oncology Reports* 18(1): 167-74.

Woodhams, R, Ramadan, S, Stanwell, P, Sakamoto, S, Hata, H, Ozaki, M, Kan, S, Inoue, Y. (2011). Diffusion-weighted imaging of the breast: principles and clinical applications. *Radiographics* 31(4): 1059-84.

Xing, RH, Mazar, A, Henkin, J, Rabbani, SA. (1997). Prevention of breast cancer growth, invasion, and metastasis by antiestrogen tamoxifen alone or in combination with urokinase inhibitor B-428. *Cancer research* 57(16): 3585-93.

Yamada, S, Ikeda, H, Yamazaki, H, Shikishima, H, Kikuchi, K, Wakisaka, A, Kasai, N, Shimotohno, K, Yoshiki, T. (1995). Cytokine-producing mammary carcinomas in transgenic rats carrying the pX gene of human T-lymphotropic virus type I. *Cancer Research* 55(12): 2524-7.

Yamanoshita, O, Ichihara, S, Hama, H, Ichihara, G, Chiba, M, Kamijima, M, Takeda, I, Nakajima, T. (2007). Chemopreventive effect of selenium-enriched Japanese radish sprout against breast cancer induced by 7,12-dimethylbenz[a]anthracene in rats. *The Tohoku Journal Of Experimental Medicine* 212(2): 191-8.

Yan, H-X, Wu, H-P, Ashton, C, Tong, C, Ying, Q-L. (2012). Rats deficient for p53 are susceptible to spontaneous and carcinogen-induced tumorigenesis. *Carcinogenesis* 33(10): 2001-5.

Yang, JH, Nakagawa, H, Tsuta, K, Tsubura, A. (2000). Influence of perinatal genistein exposure on the development of MNU-induced mammary carcinoma in female Sprague-Dawley rats. *Cancer Letters*. 149(1–2): 171-9.

Yang, N, Huang, B, Tsinkalovsky, O, Brekka, N, Zhu, HY, Leiss, L, Enger, O, Li, XG, Wang, J. (2014). A novel GFP nude rat model to investigate tumor-stroma interactions. *Cancer Cell International* 14: 541.

Yoshikawa, T, Kawaguchi, H, Umekita, Y, Souda, M, Gejima, K, Kawashima, H, Nagata, R, Yoshida, H. (2008). Effects of neonatally administered low-dose diethylstilbestrol on the induction of mammary carcinomas and dysplasias induced by 7,12-dimethylbenz [a] anthracene in female rats. *In vivo* 22(2): 207-13.

Zan, YH, Haag, JD, Chen, KS, Shepel, LA, Wigington, D, Wang, YR, Hu, R, Lopez-Guajardo, CC, Brose, HL, Porter, KI, Leonard, RA, Hitt, AA, Schommer, SL, Elegbede, AF, Gould, MN. (2003). Production of knockout rats using ENU mutagenesis and a yeast-based screening assay. *Nature Biotechnology* 21(6): 645-51.

Zhi, W, Gu, X, Qin, J, Yin, P, Sheng, X, Gao, SP, Li, Q. (2012). Solid breast lesions: clinical experience with US-guided diffuse optical tomography combined with conventional US. *Radiology* 265(2): 371-8.

Zile, MH, Welsch, CW, Welsch, MA. (1998). Effect of wheat bran fiber on the development of mammary tumors in female intact and ovariectomized rats treated with 7,12-dimethylbenz(a)anthracene and in mice with spontaneously developing mammary tumors. *International Journal of Cancer* 75(3): 439-43.

Zingue, S, Cisilotto, J, Tueche, AB, Bishayee, A, Mefegue, FA, Sandjo, LP, Magne Nde, CB, Winter, E, Michel, T, Ndinteh, DT, Awounfack, CF, Silihe, KK, Melachio Tanekou, TT, Creczynski-Pasa, TB, Njamen, D. (2016). Crateva adansonii DC, an African ethnomedicinal plant, exerts cytotoxicity in vitro and prevents experimental mammary tumorigenesis in vivo. *Journal of Ethnopharmacology* 190: 183-99.

ABOUT THE EDITORS

Antonieta Maria Alvarado Muñoz is a Graduate in Veterinary Medicine at Universidad Nacional Experimental "Francisco de Miranda", Coro, Venezuela in 2001. Magister Scientiarum in Medicine and Surgery of Small Animals from Universidad Centroccidental "Lisandro Alvarado", Cabudare, Venezuela in 2007. Doctor Degree in Veterinary Sciences from Universidade de Trás-os-Montes e Alto Douro, Vila Real, Portugal in 2017. Professor of Pathological Anatomy, Universidad Centroccidental "Lisandro Alvarado", Venezuela from 2007 to 2018, and nowadays Professor at Universidade Lusófona de Humanidades e Tecnologias, Portugal.

Ana I. Faustino is an Integrated Master in Veterinary Medicine at Universidade de Trás-os-Montes e Alto Douro, Portugal in 2012. European Doctor Degree in Veterinary Sciences from Universidade de Trás-os-Montes e Alto Douro, Portugal in 2017. Nowadays, she is Professor at Universidade Lusófona de Humanidades e Tecnologias, Portugal, and Researcher of the Centro de Investigação e Tecnologias Agroambientais e Biológicas, Portugal.

LIST OF CONTRIBUTORS

Aaron J. J. Lemus
Department of Biology, California State University, Northridge, CA USA

Ana I. Faustino
Center for the Research and Technology of Agro-Environmental and Biological Sciences (CITAB), UTAD, Vila Real, Portugal
Faculty of Veterinary Medicine, Lusophone University of Humanities and Technologies (ULHT), Lisbon, Portugal

Antonieta Alvarado
Center for the Research and Technology of Agro-Environmental and Biological Sciences (CITAB), UTAD, Vila Real, Portugal
Faculty of Veterinary Medicine, Lusophone University of Humanities and Technologies (ULHT), Lisbon, Portugal

Bianca A. Ortega
Department of Biology, California State University, Northridge, CA USA

Bruno Colaço
Center for the Research and Technology of Agro-Environmental and Biological Sciences (CITAB), UTAD, Vila Real, Portugal
Department of Zootechnics, University of Trás-os-Montes and Alto Douro (UTAD), Vila Real, Portugal

Crystal T. Lao
Department of Biology, California State University, Northridge, CA USA

Jacqueline Saenz
Department of Biology, California State University, Northridge, CA USA

Michelle Olmos
Department of Biology, California State University, Northridge, CA USA

Paula A. Oliveira
Center for the Research and Technology of Agro-Environmental and Biological Sciences (CITAB), UTAD, Vila Real, Portugal
Department of Veterinary Sciences, UTAD, Vila Real, Portugal

Randy W. Cohen
Department of Biology, California State University, Northridge, CA USA

Toni L. Uhlendorf
Department of Biology, California State University, Northridge, CA USA

Wesley M. Tierney
Department of Biology, California State University, Northridge, CA USA

INDEX

TRANSGENIC PLANTS: NEW RESEARCH

EDITOR: Oliver T. Chan

BOOK DESCRIPTION: The aim is to design plants with specific characteristics by artificial insertion of genes from other species or sometimes entirely different kingdoms. This new book presents the latest research in this growing field.

HARDCOVER ISBN: 978-1-60692-017-6
RETAIL PRICE: $135

MAMMARY GLANDS: ANATOMY, DEVELOPMENT AND DISEASES

EDITOR: Edmund B. Rucker (Department of Biology, University of Kentucky, Lexington, KY, USA)

SERIES: Veterinary Sciences and Medicine

BOOK DESCRIPTION: This book is a compilation of manuscripts that encompass a breadth of information from different species including rodents, dogs, cows, goats, and sheep. Normal developmental processes are covered and anatomical features are discussed, as well as the impact and significance of environmental chemicals and epigenetics on the mammary gland.

HARDCOVER ISBN: 978-1-62948-853-0
RETAIL PRICE: $179

CANCER METASTASIS RESEARCH: PATHOLOGICAL INSIGHT

AUTHOR: Takanori Kawaguchi (Division of Human Life Sciences, Fukushima Medical University School of Nursing, Fukushima, Japan)

SERIES: Cancer Etiology, Diagnosis and Treatments

BOOK DESCRIPTION: Hopefully, this book will enlighten people in the medical and research field regarding new research in cancer and the interested reader as well.

HARDCOVER ISBN: 978-1-61942-863-8
RETAIL PRICE: $95